MEN OF THE MACHINE

An Inside Look At Baseball's Team of the '70's

By Ritter Collett

LANDFALL PRESS, Inc.
Dayton, Ohio

Photographs courtesy of the Cincinnati Reds
and the Dayton Journal-Herald

Cover design by Robert Cyphers

Printed and bound in the United States of America by The C. J. Krehbiel Co., Cincinnati, Ohio

ISBN 0-913428-28-0

To my typist who has also legally shared my bed for 29 years and tolerated my hundreds of days pursuing fun and games on the road with the Reds.

Contents

MEN OF THE MACHINE

REDS

An Inside Look At Baseball's Team of the '70's

PROLOGUE

Baseball is a game of men . . . not machines.

Across the country, sports fans know baseball's most successful franchise of the 1970's, the Cincinnati Reds, as the Big Red Machine.

Appropriate as the nickname may be, the Reds are made of flesh and blood and muscle . . . not pistons, wires and cylinders.

This is the story of the world champions as individuals and human beings with an emphasis on characteristics from both sides of their lives — players and persons.

The earliest cogs of the machine were Peter Edward Rose and Atanasio Rigal Perez, who as hesitant and lonely teenagers spent a summer together in the metropolis of Geneva, New York, in 1960 at what was then the bottom level of the Reds' farm system.

This was one year after George Anderson, known as Sparky, put in his one and only season as a player at the major league level.

From that easily forgettable active career, Anderson is compiling a managerial record that some day surely will earn him consideration for the Hall of Fame at Cooperstown.

Is he a push button manager? What is the secret of his success?

This is the story of Johnny Bench, the little boy with Indian blood in his veins who dreamed of playing in the big leagues but whose family in rural Oklahoma was too poor to buy him a ball or a bat.

Johnny Bench has exceeded his daydreams many times over as one of the greatest catchers of all time, the No. 1 gate attraction in his sport, a figure who easily fits in with the biggest names in the entertainment world and the world of politics, and a man whose private life includes an endless stream of the country's most beautiful women.

This is the story of Joe Morgan, a talkative troublemaker by reputation with the Houston Astros who emerged as an articulate and purposeful leader in Cincinnati.

Gary Nolan, a father at 17, a big league winner at 18, and a potential has-been at 19 is a unique member of this cast of players.

David Concepcion, the flashy Venezuelan shortstop, can't stand the sight of blood, yet his father wanted him to be a doctor. Why did Pete Rose call him Bozo the Clown?

Don Gullett, the quiet Kentuckian, came out of nowhere to make the team in 1970, then became one of baseball's "instant millionaires" when he elected to depart the Cincinnati scene via the free agent draft after the 1976 season. The hook he swallowed was baited with a $2.1 million contract dropped by George Steinbrenner, the money-bags owner of the New York Yankees.

There are other cogs in the Machine, such as George Foster, the Bible-reading strongman; Rawly Eastwick, the slender, artistically inclined relief pitcher, and Fred Norman, who came to Cincinnati so his ex-employer could meet a payroll.

Dan Driessen was a catcher in a tryout camp. The Reds signed him and then decided he was too uncommunicative to relate to pitchers.

They are a diverse lot, these men of the Machine. In the superstar category are Latins, Blacks and Caucasians, plus an American Indian strain.

They come from California, Ohio, Kentucky, Oklahoma, Florida — plus Cuba, Venezuela and the Dominican Republic. Eventually the Reds acquired Michael Ken-Wai Lum, a resident of the 50th of these United States but the only major leaguer with 50 per cent Oriental bloodlines.

It is part of Sparky Anderson's genius as a manger, and a very big part of it, that he has managed to hold the reins on a team that includes so many athletes with so much built-in potential for personality and ego conflicts.

Much of the success of the Reds has been the blending of well-paid talents into an old-fashioned team concept. During the World Series of 1976, with the free agent auction only a few weeks into the future, a great many baseball people were rooting for the Reds against the Yankees because they represented baseball's traditional image so well.

"You have to have some of the little boy in you to play this game," says Joe Garagiola, the onetime catcher, who as a television personality, has far exceeded any success he had in uniform.

Garagiola once was a little boy, although presumably one without much hair. He was 19 years old in 1946 and a St. Louis Cardinal rookie when he hit his first major league home run in Cincinnati's Crosley Field. As he rounded first he could no longer restrain himself. He stopped and jumped for joy, clapping his hands before resuming his journey around the bases. When he returned to the Cardinal dugout, he got the "silent treatment." Everyone ignored him as if nothing had happened — but then the dugout burst into a roar of laughter to assure him they really were sharing his achievement.

At times the "little boy" comes through in Johnny Bench and it is definitely part of the Pete Rose mystique.

However, any pretense that Rose loves the game well enough to play it without being well paid suffers as he tools around in his $50,000 Rolls Royce, not to mention his tough salary negotiations going into the 1977 season. Yet the money was apparently important to Pete as a way of keeping score — a certain recognition of achievement.

The little boy strain in Morgan isn't all that apparent. But Joe frequently telephones his father during the season and gives him the inside of what is going on with the Reds.

Baseball history is saturated with Dennis the Menace shenanigans. Take the afternoon in Chicago's Wrigley Field which has an ancient, tiny locker room. A naked Concepcion amused himself by crawling into the clothes dryer that puts the finishing touches on the unmentionables after the game is over. Pitcher Pat Zachry sauntered along, pressed the start button and there went David, whirling around, a frightened look on his face, until a half dozen suddenly serious Reds were scrambling to open the door. The hair on his legs was singed and David isn't all that hairy.

Bernie Stowe, the only major domo of the Reds' clubhouse that any of the present cast has known, has a hundred stories about things that happened over the years since he started as a bat boy back in 1949. He's now the equipment and clubhouse manager. Some of his favorite stories deal with Frank Smith, a lanky, red-haired West Virginian who was a topflight relief pitcher in the early 1950's.

Smitty was an unrestrained practical joker. One day, after being knocked out of a game, he went to the clubhouse and took a coat hanger, fashioned a noose and waited for someone to appear. Then he leaned forward out of his locker as if he had hanged himself.

"Somebody sent one of the other kids up to the

clubhouse for something,'' Stowe recalled. ''He came running back, screaming and white as a sheet. You never saw a dugout empty so fast. There was Smitty, that thing wrapped around his neck, tongue sticking out and I have to say, he made it look realistic.''

The procedure by which the Reds took the field at that time was for the eight starters to charge out of the dugout running toward their positions. The park organist played a lively fight song until the pitcher crossed the foul line. Then the park announcer set the stage for the national anthem.

The organist's cue was seeing the players charge from the dugout. One day Smith and several of his bullpen buddies charged out — about two minutes before the actual starting time. The organist thundered into his theme, and then stopped in mid-chord, watching Smitty and friends standing a few steps in front of the dugout waving at him.

For every clubhouse prank, there is the alternate temper tantrum of a frustrated loser.

Fred Hutchinson, who managed the Reds from mid-1959 until his untimely death from cancer in November of 1964, was one of the most friendly, considerate persons who ever walked — until he lost his temper. He smashed three metal chairs into a pile of twisted junk after losing a tough game in Pittsburgh. In Crosley Field, he once threw a ball bag out of his office window — without opening the glass.

Clubhouse attendants prepare post-game food for their athlete clients. Managers have been known to tip over whole tables of cold cuts in a rage and woe to the unwary player who gets to the table too quickly after a tough setback.

Gene Mauch, longtime manager of the Phillies, later

of the Expos and directing the Twins at this writing, once had to buy three of his Philadelphia players new suits because he splashed their lockers with a tray of barbecued ribs he flung into the air. This was in Houston, once known as the Maxim's of the majors, where the original clubhouse man, Norm Gerdeman, posted a weekly menu of specialties prepared by his wife, Evelyn.

Another very human side of baseball is inevitably found in the section of the stands where the wives sit. In Cincinnati's Riverfront Stadium this is enhanced by a "family waiting room" outside the Reds' locker area where wives, kids and assorted relatives are to be found after the games.

Karolyn Rose, who has the same outgoing personality as her famous husband, is almost always to be found there, in a lively mood win or lose, although the general tenor of the room reflects the fortunes of the team. Anderson has a general rule that the kids stay out of the locker room after a defeat, but can come in after a victory.

Jolene Billingham, Carol Nolan, Robin Plummer and Pituka Perez, until Tony was traded to Montreal, were the wives who made it a habit to use the family waiting room. Others, like Gloria Morgan, do not come to the games all that often.

Wives often build more social alliances than their husbands, if for no other reason than they have more opportunities with their spouses gone at least half the time during the playing season.

Few wives are seen on the road. Baseball management does not encourage it, nor are they usually permitted to travel on team charter flights. Once a team is in the World Series, however, the wives are taken along at the expense of management. The same procedure holds true for players selected on the All-Star teams.

An understanding wife can be of great help to her baseball playing husband. She has to be aware that there are temptations with the player living on the road in something of a spotlight. She can also be of great assistance in helping him cope with the tensions of his business.

So the life of a baseball player is not all base hits and chasing fly balls. He's an individual with emotions, ambitions and problems to some degree similar to those most of you have to contend with.

This book is not a capsulization of baseball statistics, although some of these represent a part of the story.

This, then, is an attempt to provide insights into the men who have made Cincinnati's Big Red Machine the most interesting and successful franchise of its era.

Ritter Collett
May 1977
Dayton, Ohio

Hustling Rose looks for a "slip" behind him by the Dodgers. A glum Ron Cey stands by.

Karolyn, Pete, Petey and Fawn.

An Aggressive reflection

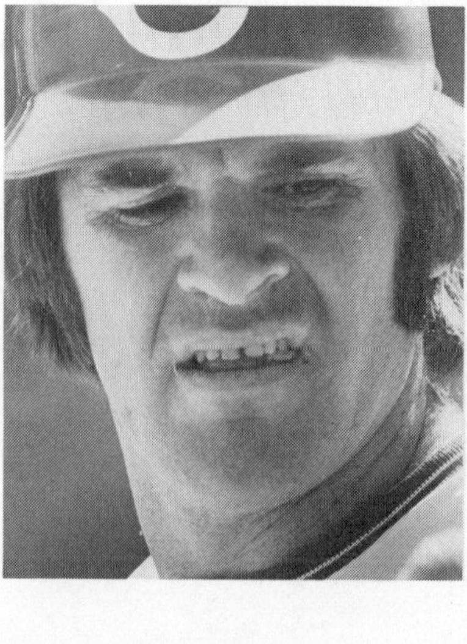

THE WHIRLWIND

It was getting close to midnight in Cincinnati's sweltering, brand new Riverfront Stadium the night of July 14, 1970 and the annual major league All-Star game was going into the 12th inning deadlocked at 4-4.

Across the country in morning newspaper sports departments, nervous editors were pleading for someone — anyone — to win the game so they could get the full story as deadlines crept dangerously close.

The Secret Service contingent guarding President Richard M. Nixon was nervous for another reason. Any change in the presidential schedule has nationwide reverberations and Air Force One was waiting to fly the President and his party back to Washington.

As the Nationals came to bat, Pete Rose, the hometown Cincinnati superstar set the crowd of 51,838 buzzing with a leadoff single and was promptly sacrificed to second. Rose took his lead in short, hopping steps typical of his concentrated intensity and was ready to break as Jim Hickman, Chicago Cubs outfielder, dropped a base hit to center. Amos Otis of the Kansas City Royals charged the ball.

To the astonishment of nearly everyone and the dismay of most, Rose churned around third base as fast and as hard as his sturdy, stocky legs could carry him.

Ray Fosse, the then 23-year-old catcher of the Cleveland Indians, took a look at Rose and knew that with a halfway decent throw the ball would be there in time for a play at the plate and, by all logic, Rose would be an easy out. But Fosse did not appreciate Rose's competitive instincts.

Fosse planted himself facing the runner with the baseball firmly tucked inside his big catcher's mitt. It never crossed his mind that Rose, whom he had met at dinner the night before, would leave his feet and barrel into him with the full impact of his 200-pound body just as Franco Harris, the great running back of the Pittsburgh Steelers, would try to run over a linebacker. There was an awesome collision and Fosse was separated from the baseball and his senses in a shock wave of pain.

Rose, knocked away from the plate by the impact, scrambled back to touch it, saw umpire Al Barlick give the "safe" sign for the winning run, and then asked Barlick if Fosse was all right. Fosse was not all right and neither was Pete, who was rushed to the training room in his home clubhouse for a quick examination while Fosse was being taken to Christ Hospital by ambulance.

The unexpected, all-out effort by Rose to win what most baseball players consider an exhibition game, is typical of Rose — and has been since he broke into baseball.

"Ray is a real nice kid and I hope he's all right," Rose told the army of reporters who crowded around him in the Cincinnati training room. "He was out to my house last night after Frank Scott (a well-known baseball agent) took several of us to dinner.

"I played to score that run the way I play 162 games in the National League, like Frank Robinson does in the American League," he told reporters.

"When I saw those shinguards of his guarding the plate, there was no way I could go under or around them and sliding into them was a greater risk as I saw it than going in straight up like I did," he explained. "I didn't have time to think about banging myself up or hurting him."

In the hospital, where X-rays displayed no fractures but a badly jammed shoulder, Fosse forced a smile.

"Damn," he said, "it's a good thing I got to know him. He's a helluva guy, but if we ever meet again on the field, I'll be better prepared."

At the time he scored the winning run in that 1970 All-Star game, Peter Edward Rose had won two consecutive batting championships and was already an established star. Going into the 1977 season, he had added another batting championship, a Most Valuable Player Award, a string of Cincinnati club offensive records and a total of 2,762 hits.

Barring an unforeseen serious injury, Rose will become the 13th man in major league history to reach the 3,000-hit plateau sometime in the 1978 season. Of the dozen who have made it, 11 are in the Hall of Fame and the other, Al Kaline, is not yet eligible, not having been retired five seasons as an active player.

One would suspect that Rose's incredible potential would have been obvious to any baseball scout worth his paycheck. But Peter Edward is the ultimate rebuttal to the notion that the computer method of evaluating human potential is a valid use of that intriguing science. If computers had been around in the 1800's, surely the machine would have low-rated Abe Lincoln's ultimate potential as county prosecutor in Springfield, Ill. — and by the same reckoning, Rose was rejected as a big league prospect.

But human scouts fell into the same trap. "Too small," most wrote. "No power," still others added. "He'll never throw anybody out with that arm," was another judgment.

Rose was small when he graduated from Cincinnati's

Western Hills High School in June of 1960, one year behind the class he started with.

"I weighed maybe 155," he recalls. "I had a lot of trouble in school, particularly my sophomore year, and it took me a fifth year to get through."

There were three scouts, all based in Dayton, who recognized that Pete had a future. One was Jack Baker, who still represents the Baltimore Orioles in southwest Ohio, and another was a fellow Rose called "Uncle Buddy."

Ed Bloebaum was indeed Pete's uncle, his mother's brother. A one-time minor leaguer, a longtime semi-pro player and teammate of Pete Rose Sr., he worked at the National Cash Register Company in Dayton and was a "birddog" scout for the Reds, meaning a part-timer who worked with a full-time scout, in this case the veteran Carl Ackerman. It was Buddy Bloebaum who placed Rose with an amateur team in Dayton's fast-paced summer circuit in 1959, the year he should have graduated from high school.

"Uncle Buddy convinced Carl (Ackerman) that I was going to grow heavier and stronger, he knew the history of our family. I was up to 190 in three years when I had my chance to make the Reds," Rose explained.

What scouts who came from elsewhere with no knowledge of Pete's background could not assess was the intensity of his competitive instincts as well as a disregard for danger and an astonishing level for pain tolerance.

Pete Rose, the little boy, thought it was natural to do everything with a whirlwind flourish because his father, whose given name was Harry but who was known as Pete, did things that way. Young Pete, like his father, grew up a few miles west of downtown Cincinnati in an area known as Anderson's Ferry. An actual ancient ferry boat still crosses from there to the Kentucky side, providing, among other things, a "short cut" to the Greater Cincinnati Airport from the west side of the Queen City.

The hills come close to the river bank and Anderson's Ferry is only two streets wide. But there are oil storage tanks along the river, a generous assortment of railroad tracks, a well-travelled highway and in Pete's early boyhood there were street car tracks.

"My Dad used to talk of hopping freight trains as a kid, swimming across the river, riding the waves behind steamboats in a johnboat and all kinds of things," Rose said in talking of his beginnings.

"We used to do those things and he'd act horrified when I told him," the energetic athlete continued. "Scooting in a johnboat behind a paddle wheel steamer is a wilder ride than any roller coaster. You drop in the swell so low you can't see the hilltops on either side of the river and then you ride the top of the wave like you're taking off into the sky."

Pete's affinity for sports is inherited. His father was still playing semi-pro football at age 42 and is legendary in amateur circles in Cincinnati as a guy who literally would run into a brick wall to catch a ball or tackle a runner. The elder Rose, who worked for one of the city's bigger banks, wore his hair in a rugged crew cut all his life, the same style his son wore until he went "mod" with the stylist hair-dos of the 1970's.

"My Dad was a great athlete," Pete insists to this day. "In summer I was the bat boy for his baseball team and the water boy for his football team in the fall. It never occurred to me that he was usually the oldest guy playing. He never did anything slow.

"Those guys would play on Sunday afternoons and then sit around the Trolley Tavern in Anderson's Ferry if it was a home game or a tavern if it was across the river or somewhere else and have a few beers. One year, when I was seven, I sneaked some beer over in Kentucky. I sneaked too much because, man, I threw up in two states. Maybe that's why I've never liked beer. Dad was never much of a drinker either. But he'd like a beer or two in later years when he was watching sports on television."

The close relationship between father and son was perhaps influenced by the fact young Pete had two older sisters and he was the first boy. A younger brother, Dave, came along five years later.

"I was always throwing a rubber ball against the big brick wall of Schulte's Fish House," Pete says. "wherever I went, camping or playing, I always had my ball glove strapped onto my belt."

In 1950 when Pete was nine years old, his father enrolled him in a Knot Hole league program with the stipulation that the manager permit the boy to become a switch hitter, ignoring the fact that he was naturally right-handed. "He might be a little weak left-handed for a while," said Rose the elder. "But he's got a lot to learn and he'll get better."

It was also decreed by the elder Rose that young Pete would start out as a catcher. "There are never enough catchers to go around in the big leagues," he told his son.

Once he started into organized baseball, Pete was destined never to be competitively close to his kid brother. Dave, outwardly as aggressive as Pete, ultimately was signed by the Reds but dropped out of the game after two disappointing seasons in the low minors.

It took Pete five years to get through Cincinnati's Western Hills High School, mainly, he admits, because he was "goofing off" so much of the time. In his own mind, Rose is convinced that the fact he was not invited out for football his sophomore year was a turning point that could have been disastrous.

"I had a great time my first year and was on the football team, even though I was so small the coach said he was afraid I'd get killed. When I went back to school without being in football, I started doing a lot of dumb things and nearly drove my poor mother and dad wild.

"To put it simply, I flunked my sophomore year. I had spent too much time walking out at noon and not going back. My Dad read the riot act to me so many times it just lost its meaning. Somehow, after it happened and I knew I had failed, it sank in on me that I was wrong, dead wrong. I was letting my family down and myself too. Naturally, I played baseball that summer and realized I had to prove to everyone I could make the grades and maybe play football again.

"So when I went back, I hit the books and while I was never what you'd call a brain, I was putting out to the best of my ability and I guess the teachers realized that."

He also was growing physically and found his way onto the football team again. During the fall of 1958, in what should have been his senior year, Rose was the mainstay on one of the best teams Western Hills High School ever had.

He was no longer eligible for high school sports during his fifth year, the 1959-60 term, but he had gotten his foot inside the door of Crosley Field. One of the "characters" around the park of that era was lanky John Brosnan, the "bird dog" scout for Buzz Boyle, the ex-big leaguer who was then Cincinnati's resident scout. (John's son, Jim, a pitcher with the Chicago Cubs, was destined to be a part of the Reds' 1961 championship team).

One of Brosnan's "duties" for Boyle was rounding up teen-agers who would shag fly balls during batting practice sessions with the Reds. Young Pete got involved in that.

"My Dad had taken me to a hundred games over the years, and I never missed a Reds game on radio," Pete recalled. "So it was a heckuva thrill to be in the park, getting into a uniform. Of course, we kids were out of uniform before the game started, but we were in the park and we stayed and watched the games. I remember seeing Fred Hutchinson, the manager, but I never really met him. He'd say 'hello' but that was all. Who would have thought a few years later, he'd be the man who gave me my big chance?"

After Pete finally graduated in 1960, Bloebaum, supported by Ackerman, persuaded Phil Seghi that his nephew was worth the investment of a minor league contract. Seghi, current general manager of the Cleveland Indians, was then farm director of the Reds. The day the contract was offered, Pete's father urged his son to hold off because Baker had promised an offer from the Orioles. The boy was impatient, however, and signed for $7,000 with a promised bonus of $5,000 if and when he made the major league club.

"I was so impatient," he recalls, "that when Mr. Seghi suggested I could even wait until the next season to report to a minor league team (because the season was then more than a month old) I thought they might change their minds."

Rose signed on Saturday and on Monday flew to Geneva, New York, where the manager shifted a young Cuban from second to third base to make room for the newcomer. The shifted player was Tony Perez.

Life in the low minors was a rude awakening for Pete, as it is with most starry-eyed youngsters who dream of a baseball career, starting with a daily routine of rooming houses, hamburgers, long bus rides and tiny, often unsanitary, locker rooms.

"It was rough," Pete recalls, "I lived in a private home with three other guys. The Cubans, most of whom spoke little English, had it a lot worse."

One of the Cubans who preceeded Rose and Perez into Geneva was the fun-loving Chico Ruiz, who died in an automobile accident in 1971 when he was with the San Diego Padres.

"Chico used to tell about going into this restaurant in Geneva," Rose tells the story, "and he was determined to order off the menu without help. He pointed to three different things and the girl brought him three different kinds of potatoes."

Rose didn't tarry long in Geneva or his other two minor league stops at Tampa in 1961 and Macon in 1962. The year in Macon was a fateful one in that Rose began his association with Dave Bristol, who was working his way up the ladder in the Cincinnati system as a manager.

The irrepressible Rose, never one to take boredom lying down, posed some unusual problems for his minor league managers. One night in a hotel in Augusta, Pete and two teammates, one who must remain anonymous, and outfielder Art Shamsky, heard an interesting "party" going on in an adjacent room. To satisfy their urge to look in on it, Rose suggested they climb out on the roof of the four-story building. Rose and Shamsky made it into position to spy on the unsuspecting "party" but the anonymous pitcher stepped on a fire escape transom opening and tumbled through to the floor below. Fortunately, he was uninjured.

The team travelled in station wagons. "Things got crowded with eight guys in wagons that weren't air-conditioned," Pete recalled. "This one hot night, I climbed out the back window and stretched out on the equipment bags piled on top. I cooled myself off and decided I'd peek over the windshield and give the guy driving a little thrill. He like to killed all of us."

That act took place as the wagon was travelling at 60 miles an hour over a two-lane Georgia highway.

Despite Rose's impressive credentials (a .330 batting average at Macon and leading his league in triples at both Tampa and Macon), he was not looked upon as likely to stay with the team when he was invited to spring training with the Reds in 1963. But Fred Hutchinson, who had managed the Reds to a surprise pennant in 1961, was convinced he was going to give Rose a thorough look as a candidate to replace Don Blasingame at second base.

"All the veterans on the Reds were convinced I was a real hot dog that spring even though I tried to tone myself down," Rose said in retrospect.

Indeed that was a majority sentiment and many of the veterans were shocked when Rose was signed to his first major league contract a few days before opening game in Crosley Field. That situation created a problem for Rose, who was forced to turn to the black players for friendship — a move that bothered Bill DeWitt, the owner, who did not comprehend the antagonism to Rose by friends of the likeable Blasingame.

Rose made his major league debut April 8, six days short of his 22nd birthday. There is more than a dash of irony in the fact that in his first time at bat he drew a base on balls off Earl Francis, journeyman Pittsburgh Pirate pitcher, and then went hitless in three more at bats. On that leadoff walk, Pete threw his bat away and broke for first base on nearly a dead run, and Charlie Hustle was born. Running to first base on a walk is still one of his trademarks and a reason people either love him as a hustler or hate him as a hot dog.

Rose is proud that the first of his many major league hits was a line drive "gapper" off Bob Friend, one of the league's premier pitchers. It was in the gap between the outfielders and Pete hustled it into a triple.

Despite the fact he was playing an energetic second base with more range than the man he displaced, the resentment of his white teammates intensified for a while. Frank Robinson and Vada Pinson became his close friends. On the occasion of his first visit to Chicago, he was out after curfew (only a few minutes, he insists) and upon his return discovered his roommate had the safety lock on the door and ignored his knocks and appeals to be admitted.

"I didn't know what to do, so I went to Vada's room and he took me in for the night," he recalls.

The pitcher who played that shabby trick hardly deserves anonymity and was on the Cincy scene less than a full season.

If some of his teammates were slow to accept Rose, Cincinnati fans were not. His hustle, his good-natured brashness and his aggressive bat were making friends. He won "Rookie of the Year" honors with a .273 batting average over 157 games — and promptly was called to active duty by the military reserve unit to which he was attached.

On Jan. 25, 1964, on a weekend pass from Fort Campbell, Ky., Rose and Karolyn Ann Engelhardt were married following a whirlwind courtship. That evening Pete, accompanied by outfielder Shamsky, his best man, dashed from his own wedding reception to a downtown hotel to pick up a trophy from the Cincinnati baseball writers.

Over the years, Karolyn Rose has become a celebrity

in her own right. With a personality as outgoing and energetic as her husband's, Karolyn has been justifiably called the "perfect mate" for the human whirlwind she married.

"I met him at the River Downs Race Track and I had no idea he was with the Reds," she explained. "I didn't know that first thing about baseball."

"That's the truth," Rose interjected. "She asked me what Maury Wills was doing with all those bases he stole. She thought he was taking them home with him."

Nearly everyone needs a stabilizing experience before he solidifies himself and Rose was no exception. The 1964 season was his "shock probation" and, happily, he's never needed another.

"I thought I was making progress but I was told I wasn't and they were right," Pete concedes. "I wasn't studying the pitchers and bearing down on my hitting."

Dick Sisler, acting manager during the period Hutchinson was dying of cancer, even benched Rose briefly. He was asked to play winter ball in Venezuela after that unhappy season and said the experience did him a world of good.

The 1965 season was the first of nine straight in which Rose was to hit over .300 and start his climb toward super-stardom. He hit a robust .312 as the Reds enjoyed a super offensive season but never quite caught up with the pennant-winning Dodgers mostly because of pitching deficiencies. Rose wound up with 209 hits, the first time he had led the National League in any category.

Two off-season jolts followed Pete's break-through as a .300 hitter. Owner Bill DeWitt, who had never really wanted Sisler as a manger, fired him during the World Series. Dick had been Pete's first batting tutor at the big league level and they had enjoyed a smooth relationship after Sisler disciplined him during the '64 season.

Don Heffner, the new manager, announced that Rose would be shifted to third base to make room for Tommy Helms at second. Helms and Rose had roomed together at Macon and would do so again in the majors. Even so, the move did not sit well with Rose, who had to read about it in the papers.

In December DeWitt shocked the baseball world by trading Robinson to the Baltimore Orioles. Reached for comment on the trade of his friend, Rose blurted, "they've gotten rid of the best ball player they ever had." And he was right — until he himself disproved the point.

The first half of the 1966 season was an unmitigated disaster for everyone with the Reds. The team got off to a terrible start, morale sagged and Rose had a mental block about playing third base. After 16 games, the harrassed Heffner shifted him back to second and sent Helms to third. Pete pulled himself together and finished strong for a .313 average.

Midway through the year, the unfortunate Heffner was fired and Bristol named interim manager. It was a fateful decision for Rose because Bristol had come to the conclusion that Pete's future was in the outfield and convinced him to make the switch. Approached by a man in whom he had confidence and knowing his move would help Helms, Pete attacked his new position in the spring with enthusiasm.

The '67 season started off with the Reds floating on clouds 10 and 11 as well as nine. Out in front of the pack, the Reds didn't fall behind the Dodgers until late June. But there was a reason — in the same game in Dodger Stadium the club lost shortstop Leo Cardenas with a broken wrist and Rose fractured his collar bone trying to make a diving catch.

Incredibly, Pete missed only 14 games but his batting suffered with the restrained swing and he was fortunate to keep his .300 string alive, doing it with .301. But consecutive batting championships were just ahead and the incomparable Rose won the crown in 1968 despite suffering a fractured thumb, again in a mishap in Dodger Stadium.

This time he was on the disabled list for 21 days, a move that Karolyn Rose found put a great strain on the tranquility of their domestic life.

"If it ever happens again, don't put him on the disabled list, put him in a cage," she told skipper Bristol.

The race for the batting title was up for grabs between Rose and Pittsburgh's tiny Matty Alou going into the final weekend of the season. The San Francisco Giants were in Cincinnati and on Saturday Gaylord Perry — one of Rose's few jinx pitchers — was due to work. All he did was go four-for-five against Perry, whose reputation for throwing a "wet ball" was well known.

"I'll never know how I did it," Rose said later. "Every hit I got was off a Vaseline ball."

The following season, with his salary pumped up to $85,000 via a noisy spring holdout, Rose repeated, making off with the batting title this time against the late, great Roberto Clemente. Again the issue was in doubt down to the wire and the last of Rose's 218 hits was a bunt single in the eighth inning in Atlanta.

The 1970 season was a high water mark for Rose in every way but a third batting title. He achieved his goal of becoming baseball's "first $100,000 singles hitter" — something he had been talking about for three years.

The unknown Sparky Anderson had taken over as manager and Rose, who had been named team captain by Bristol, volunteered full cooperation with the new manager in helping him get his feet on the ground.

The Reds moved from Crosley Field where Rose had watched so many games as a boy into new Riverfront Stadium where he won the All-Star game with his smashing arrival at home plate.

The Reds waltzed to the western division championship and for the first time Rose found himself in the excitement of post-season play. He was not a key figure as the Reds captured their first pennant, sweeping a three-game series with the Pirates in which pitching surprisingly dominated two heavy-hitting teams. But then the Baltimore Orioles blitzed the Reds in a five-game World Series in which Pete had a home run among his five hits.

1970 is also the year during which Rose became acquainted with his son. Peter Edward Rose II was born November 16, 1969, nearly five years after his sister, Fawn, the first Rose youngster. But if his infant son motivated him during the 1970 pennant drive, the year was to end on a tragic note. His father died of a heart attack in December at age 59.

"He died much too soon," said Pete sadly at the time. "But thank God, he lived long enough to see me in a World Series and get to know his grandson."

In vastly different surroundings, the Reds' star concedes he is trying to establish with his son the same sort of relationship he enjoyed with his father.

Petey, sometimes known as "Googey," has been a familiar figure around the Cincinnati clubhouse almost since he learned to walk. His father bought him plastic bats and balls which he attacked with reckless gusto. On one occasion, he created a stir with relief pitching ace Clay Carroll tossing him a plastic ball. "Get that bleep over," he shouted, repeating a phrase he had heard his father and other Reds yell at pitchers in batting practice on the field.

"This is the most natural place in the world for him to be if he's going to spend time with me," says Rose. Young Petey was the bat boy at exhibition games in Tampa in the spring of 1977. One day Johnny Bench was teasing him about something and was startled to hear the kid say, "Betcha 20 bucks I'm right."

"I was 19 years old before I ever saw a $20 bill," said Bench to a bystander.

The life style in which Peter Edward II is growing up is vastly different than his father's was, although they still reside in the same general area. The difference is that instead of down along the railroad tracks, the Roses live in a sprawling house up in the hills overlooking the river. The house belonged to the late William H. Zimmer, once a stockholder in the Reds and an official of the Cincinnati Gas and Electric Company. Bill was the uncle of Don Zimmer, the Cincinnatian who came out of Western Hills High School a decade ahead of Pete to also become a major leaguer. He was named manager of the Boston Red Sox in the middle of the 1976 season.

It has occurred to Rose that eventually his son might find other interests stronger than sports.

"I don't know," he said, musing on that question. "Sports was so much a part of my dad's life and so much a part of mine, I hope he stays with it. But we'll have to see."

Daughter Fawn is inevitably in attendance at Riverfront with her mother.

Rose looks back on 1972 with bitter feelings in that the Players Association called a strike late in spring training which delayed the start of the regular season and cost the Reds eight games that were never played. What it meant to Rose was that he finished the year with 198 hits — two short of his perennial goal of 200.

Only one man in baseball history has had nine seasons in which he had 200 or more hits — the legendary Ty Cobb, considered by many the best hitter who ever lived. When Rose amassed 215 hits in 1976 it raised his total to eight seasons in that category. He knows the strike cost him his ninth year.

Anderson had asked Rose to shift over to left field from right going into 1972, making it the fourth position at which he had opened the season for the Reds. The Reds led the Western Division race from June on and once again faced the Pirates in a playoff. Unlike the 1970 sweep, this time the Reds fought an uphill struggle through the fifth game when Johnny Bench's dramatic home run turned the tide for a thrill-packed 4-3 victory.

The '72 season was notable for bringing Joe Morgan into the Cincinnati cast and adding a considerable luster to an already impressive array of talent.

Rose and mates were bitterly disappointed in the World Series as the Oakland A's outlasted them in seven tense games. Pete had his finest hour in game five in Oakland where Charles O. Finley's team had an opportunity to finish the Reds off and prevent the return of the Series to Ohio. After arousing the wrath of Oakland fans by saying pitcher Catfish Hunter did not overly impress him, he led off that game with a home run off Hunter and then drove in the winning run in the ninth inning of the 5-4 game with a base hit off relief ace Rollie Fingers.

Karolyn, for reasons known only to herself, sat in the left field seats in the Oakland Stadium in game four amidst the people booing and throwing things at her husband.

"One Oakland fan must own a vegetable market from all the stuff he and his friends threw at me," Pete quipped to a newsman.

The longest 90 feet Rose never was able to negotiate

came in the eighth inning of the deciding game when the Reds left him — the potential tying run — stranded there in the course of a 3-2 setback.

The 1973 season sent the human whirlwind into the second decade of his brilliant career and in many ways was a climax. He captured his third batting championship with a robust .338 and played a key role as the Reds overcame a whopping deficit to run away from the archrival Dodgers in the Western Division race. And Rose was voted the MVP in November, joining teammate Bench who has won that award in the pennant-winning campaigns of 1970 and 1972.

Yet there was the infamous "Battle of Shea Stadium" during the bitterly contested playoff with the New York Mets, who had won the NL East with 82 victories compared to the 99 taken by the Reds in the other division.

The "Battle" was triggered when Rose, with his team behind 9-2, "attacked" shortstop Bud Harrelson and wrestled to the ground the slender New Yorker he outweighed by 40 pounds. The scuffle finally was broken up but when Rose returned to his position in left field a shower of debris, including at least one whiskey bottle, was fired at him. Anderson pulled the Reds off the field until calm was restored.

"I did it trying to wake my team up," Rose said later. "If I had wanted to punch Harrelson out, I could have done it. He gave me the opportunity to grab him when I slid into him and he said something to me. I didn't want to hurt him. I just tried to stir something up."

Later that winter, at the annual dinner of the New York Baseball Writers, Rose was given the chapter's "Good Guy Award" and he planted a kiss on Harrelson's cheek as the two posed for photographers.

The loss to the Mets stirred Rose and the rest of the Reds to yearn more than ever for a coveted world championship. In spring training, Bench coined a phrase, "A bridesmaid no more in '74," but it was not to be. The club never caught the Dodgers in the west and Rose's streak of .300 seasons came to an end when he hit a mere .284.

The whispers that he was fading into the sunset built a strong, new competitive fire in Rose going into the fateful 1975 campaign. For the first time, he talked about the passing years and the need to bear down harder.

"I always make sure our new kids get to spend a day with the captain early in training camp," Manager Anderson says. "They see him put on his long johns, then a rubber suit, then a heavy shirt and go out and wear out a coach taking ground balls and then stay for extra batting practice."

Pete asked more of his rock-hard body in the spring of '75 than ever before. The 155-pounder who graduated from high school has played at weights as high as 208 without losing much efficiency and without revealing a trace of fat. Rose's torso looks like it was clawed out of a mine shaft. He is not as muscular as he is solid. You might be able to strike a match on his chin or his ankles. His pain tolerance is fantastic.

"There's no way a man can be too tired to play baseball," he insists with what to him is infallible logic.

Reds team physician, Dr. George Ballou, recalls when Rose had his broken thumb back in 1968, Pete kept insisting the cast should come off so he could play. "The day I did take the cast off, I didn't think he could swing a bat for at least a week," Dr. Ballou said. "I gave him a wet towel in my office and told him to slam the wall with it as hard as he could. I expected to see him cringe in pain. He never let out a whimper."

The start of the 1975 campaign found the Reds spinning their wheels through April. On Friday evening of May 1, Rose was taking some ground balls at third base when Anderson called him over.

"I'd like you to take a crack at third base," the manager said. "I know it's asking a lot, but we've got a problem there. If you move in from the outfield, I can platoon George Foster and Danny Dreissen out there and I know it will help the club."

"When do you want me to start?" responded Pete.

"Tomorrow," answered Sparky.

Tomorrow it was, and the Big Red Machine was on its way to the long-awaited world championship that Rose and mates wanted so much.

"I know Pete can get by over there," Anderson explained. "He can do it on hustle alone."

Foster disrupted Sparky's plan for a platoon in the outfield spot by coming into his own as a consistent hitter and the producer of some outer space home runs. George drove in 78 runs, while Rose led the league in doubles, wound up with a .317 average and enjoyed a super playoff series against the Pirates — victimized a third time since 1970 — and then against the Boston Red Sox in the dramatic 1975 World Series.

What the world didn't know was that Rose was taking cortisone shots in his throwing arm much of the season, the soreness caused by his extra efforts to make himself a better third baseman.

With the Red Sox ahead 3-0 in the seventh Series game, Rose led off the sixth inning with a single and, although he himself was erased in a forceout, his hit was good for a run when Perez unloaded a home run to cut the edge to 3-2. Rose's single in the seventh tied the game and then the Reds won it in the ninth.

Peter Edward finally had his World Series diamond ring in his 13th season in the majors.

Amazingly, he was within reach of another batting championship in 1976 but never quite made it. The fact a 35-year-old athlete comes that close isn't routine procedure. Rose was willing to settle for a .323 average, a strong showing in a playoff sweep over the Phillies and the "sweet" four game sweep over the New York Yankees. His 215 hits for a total of 2,762 put him within reach of Frankie Frisch's 2,880, the most ever by a switch-hitter.

Years ago Vin Scully, the talented voice of the Dodgers, quipped, "Pete Rose just beat out another walk."

He gets his share of those, but it is base hits that make up the Pete Rose legend and he's piling up enough of 'em that one of these days he'll bounce into the Hall of Fame.

Anyone care to bet he slides in head first?

Griffey's classic bunting form

Safe against Mets' Felix Millan

THE LONG SHOT

Of all the great individual talents that blend into Cincinnati's championship team, none has a greater reservoir of untapped potential than lightning-quick Ken Griffey. The man who yielded the 1976 batting crown to Chicago's Bill Madlock on the final day of the season is viewed as being on the threshold of super-stardom.

For all of that, Griffey's emergence to that level is comparable to a 90 to 1 shot winning the Kentucky Derby. This speedster out of Donora, Pennsylvania, was almost an after-thought in the 1969 free agent draft. He was taken in the 29th round, as more than 500 young players were placed ahead of him on major league scouting lists.

The Reds didn't even invest in a long distance call to negotiate with Ken. They sent him a letter and a contract, the message starting off with congratulations and then an offer to report to the club's rookie league team in far-off Bradenton, Florida.

"Hey, Mom, I got drafted," Griffey yelled to his mother upon opening the letter. Inasmuch as the Vietnam war was still very prominent in the daily headlines, Mrs. Griffey's reaction to her son's idea of a joke was a shock wave of concern.

In 1969 Griffey had not yet developed into the full-muscled 200-pound specimen he is today.

"The fact is, I was a skinny 160-pounder," he laughs in looking back to his "Roots" as a baseball player.

He had been a running back in his high school in Donora, a grimy mill town along the Monongehela River southeast of Pittsburgh. From the same town a young man named Stanley Musial emerged in the late 1930's to become a baseball immortal and a millionaire.

"I played football for three seasons and I was hurt all the time," Ken said. "I didn't care for football that much, but it looked like I was going to have to accept a college scholarship to get anywhere in the world."

The memory of his broken wrist, bruised knee and pulled hamstring induced Griffey to take a whirl at baseball and the Cincinnati offer.

"We drafted Kenny because we knew he could run," says Sheldon (Chief) Bender, the director of player personnel of the Reds. "He actually had very little baseball playing experience. When he reported to Bradenton, it didn't appear that he would ever be a good contact hitter and his throwing arm wasn't impressive. All he had was that speed and a likeable disposition."

Griffey chuckles in disputing Bender's opinion that he had limited baseball experience.

"We played baseball all the time when I was a kid," he said. "Of course, it wasn't HIS kind of baseball. Our diamond was on the playground of Highland Terrace. That's the housing project where we lived.

"We used bricks for bases and a piece of cardboard for home plate. You'd slide into it and it would slide away from you. We used powdered milk we'd sneak out of the kitchen to mark base lines. If you hit the ball good to left center it would wind up over in the swings and see-saws and be a home run."

The playground at Highland Terrace is no more. A concrete highway runs through the area, displacing most of the battered housing project.

"I feel a little sad when I go back there and see it today," Griffey confesses. "Those were the good old days and baseball in a way was as much fun then as it is now when I'm making good money playing it.''

Griffey was a modest success at Bradenton. In September he went back to Donora and married his sweetheart, Alberta Littleton.

The nightmare year in his life was just ahead. He was assigned to play at Sioux Falls, that South Dakota community where pitchers Gary Nolan and Don Gullett made spectacular debuts in the Cincinnati system.

Griffey's debut, however, was far from spectacular. ''If it hadn't been for the patience of Russ Nixon, I would never have made it,'' he confesses.

Nixon, who joined the Reds coaching staff in 1976, is a onetime American League catcher who displayed a rare talent for communication with uncertain youngsters in the low minors — a quality which led the Reds to pay him ''over scale'' managerial wages to stay in the lower level of the system until he was promoted all the way to the big club. Nixon's patience with Griffey, coupled with Reds' President Bob Howsam's theory of looking beyond the physical ability of young players, saved for the Cincinnati organization the services of a player who likely would have been released and forgotten by many other baseball farm systems.

What happened to Griffey in one game in Sioux Falls defies any and all logic. He can laugh in telling the story today.

''We were in the ninth inning with a 6 to 4 lead. The other team had two men on base, two out and the batter hit a fly ball to deep left. It was high enough for an easy catch. But the ball hit off the heel of my glove, bounced over the fence for a home run and we lost the game 7-6.

''Our pitcher was a kid named Billy Wall and he quit the team and walked out of baseball the next day. Russ

(Nixon) had this dazed look on his face. If I tried to let a ball bounce off my glove over the fence I couldn't do it once in a thousand times,'' Griffey recalled.

But it was off the field where Ken's mind-blowing frustration nearly ended his career before it began. The problem was that blacks are scarce in South Dakota, a culture shock not many 19-year-old youths out of a black environment are prepared to deal with.

''All of a sudden, a lot of weird things were happening,'' he said. ''I had brought my wife up in June and our first little boy was only a couple of months old. We'd be walking down the street and cars would stop and people would be looking at us. I thought we must have been the only black family in all of South Dakota. I was highly insulted but then I'd realize maybe they'd never seen a black family and meant no harm or offense.

''I was the only married guy on the team but there were two other black players. We couldn't find places to live. I was paying $9 a day in a motel and I was only making $450 a month. We weren't eating too good.

''There was a road trip coming up and I wasn't going to leave my family there and go away. I told Russ unless we found an apartment I was going to have to give it up. It was my wife's first time away from home and I wasn't going to let her stay where people were looking funny at us.

''Finally, the guy who owned the motel said he'd find us a place to stay and he did find us an apartment for $65 a month. That was a load off my mind, but so many things were eating at me that I didn't have a good season,'' he concluded.

In spring training the following year the same players who had done better than Griffey at Sioux Falls were available for a promotion to Tampa where Nixon was to start five full seasons as manager.

"There was some discussion about cutting Griffey," Nixon recalled. "But Shergs (George Scherger) and I were against it. I said I'd be glad to have him."

Griffey himself was afraid he might be cut and concedes that Russ "went out on a limb to keep me." And his development at Tampa is an excellent vindication of Howsam's three-year theory regarding youngsters.

"One year is not enough," he maintains. "If we sign a young man, we should have a three-year look at him before we give up if we feel there is any real potential there, and we must have felt there was some if we signed him in the first place."

A heavier, stronger and more relaxed Griffey batted .342 on the season (98 points more than he had hit the year before) and began to show much more coordination.

"I know it sounds crazy," Nixon said in spring training of 1977, "but he was running better than he had before. Strength can have something to do with speed."

There still was concern by some in the Reds' farm system about Griffey's batting style. Unorthodox batting could be endemic to Donora residents, as evidenced by Musial's unique corkscrew posture at the plate. In Griffey's case, it wasn't posture. It was what appeared to be a lazy, slow swing. That is not a style normally guaranteed to make contact with the ball.

"Kenny's swing is a big sweeping thing that misleads you," says Ted Kluszewski, the man mountain batting instructor of the Big Red Machine. "His big problem was keeping two hands on the bat. We kept constantly hounding him to hang on with that top hand, which gives him better speed. The other hand sweeps, but he does make contact and he's strong at the end of the sweep."

Big Klu should be a specialist on the sweeping bat

swing. His wrists, about the size of a small tree trunk, were seldom bent as he took his cut. "Arm hitters" are usually poor hitters, unless the arms have the massive strength of a Kluszewski.

The question of Griffey's allegedly lazy bat was never fully resolved in the minds of some of the Cincinnati brain trust until he proved he could hit big league pitching.

By the spring of 1974, Griffey was a very impatient young man. The Reds had made a deal with the Baltimore Orioles to obtain the veteran Merv Rettenmund, which Griffey correctly interpreted cut down on his chances to finally make the "big" team. The year before he had batted .327 at Indianapolis, the top farm team, and had led the American Association in stolen bases. He felt five full seasons in the minors was adequate development.

As the time to cut the squad approached, a decision was made on Griffey — he would go back to Indianapolis for the time being.

"I told him that if he went back and did the job one more time, he would be back," manager Anderson recalled. "The problem was that we had sent George Foster back to Indianapolis the year before on the same promise and he was back to stay.

"I know that trip back did Foster a world of good and it also did Kenny a world of good. They were working that much harder and they appreciated being in the majors that much more."

Another factor in both Foster and Griffey's acceptance of what seemed an unnecessary and cruel delay in their careers was the steadying influence of an ex-pitcher named Brooks Lawrence, an unsung member of the front office staff.

Brooks and Dolly Lawrence are a perceptive pair who have touched the lives of many young players in the

Cincinnati system, both black and white. When Brooks broke into pro ball in the early 1950's, legal segregation was in effect in the South. In their travels around to the minor league teams, the Lawrences have provided valuable insight to help youngsters cope with frustrations.

Anderson was true to his word and Griffey was back in Cincinnati in time to play 88 games of the '74 season's frustrating chase of the Dodgers. Ken batted .251 and began to get a feel against major league pitching. Going into the 1977 season, Griffey conceded that the fact the Reds did not push him had helped his progress.

"When you come up to a team and see people like Joe Morgan, Johnny Bench, Tony Perez, Pete Rose, you have to be a little awed," Griffey says. "You ask yourself, 'what am I doing here?' But I remember what my older brother Bill used to say all the time, 'Kenny, when you walk out there to do your thing, remember you're as good as anybody else' and I've always felt that way.

"But the big thing about the Reds is that those same people — Morgan, Perez and others — they make you feel like you belong. They go out of their way to help."

Griffey's blazing speed is guaranteed to produce a number of infield hits, and on occasions he has been known to resent some of the clubhouse teasing he was taking on that score. But there's no doubt that speed helped Griffey achieve his .305 average in his first full season as he played a role in the surge to the World Series triumph over Boston in '75. Of his 141 hits, 38 were of the infield variety but only seven were bunts.

"He doesn't have to bunt," skipper Anderson says. "He tops a ball to the right side and forces everybody to rush their movements and he'll often beat the play. His speed brings the third baseman up so far that he'll slap a few ground balls through the left side for hits that would be routine outs for the average batter.''

Anderson was taking Griffey out against certain types of lefthanded pitching for awhile, but relented on that.

By the time the championship series with Pittsburgh rolled around, Griffey was ready for a strong showing against what could be called his hometown team. He had a two-run double in the 8-3 first game victory and his bunt leadoff single in the 10th inning of the third and final game enabled him to carry the winning run across the plate on Ed Armbrister's sacrifice fly.

It was Griffey's double in the ninth inning of the second game against Boston in the World Series that accounted for the winning run that enabled the Reds to come out of Boston even after two games.

The Griffey timetable for 1976 called for him to begin becoming an intelligent base runner and the heir apparent to Joe Morgan as the club's top base stealer.

''Joe has been very helpful in talking to me about how to analyze pitchers, but nobody can tell you when to take off to steal a base,'' Ken explains. ''What Joe does is study how quickly and how often a pitcher comes to first base and how quickly he goes to the plate. It's not an easy knowledge to acquire, and a lot of it has to be instinctive. The decision on when to go and when not to go has to be entirely your own.

''I don't ever see my potential stealing 100 bases with the Reds. Joe steals when it helps the team, and you just don't steal to be stealing. If I'm on first base and we need a run to tie, it's my job to get to second base.''

It is ironic that in a year the emphasis was on developing Griffey's base running talents that he nearly made off with the batting title. The Reds had long since wrapped up the Western Division title by the time the meaningless last game with Atlanta rolled around. Griffey had what amounted to a four point lead (.337 to .333) over the Chicago third baseman.

Griffey's name was posted in the original batting order when Morgan called Ken aside and suggested that it might be wise to sit on the lead in the dugout. Manager Anderson, who later insisted the decision to take Griffey out of the lineup was his and his alone, had gone into a series of discussions with Morgan, Perez, Bench and Rose prior to the game.

Then word came through from Chicago during the afternoon that Madlock was having a perfect day and Griffey was rushed into the game in Riverfront. He struck out twice. Madlock finished two points ahead (.338 to .336) for a second straight crown.

It was a strange scene in the Reds' clubhouse with no one willing to talk about the day's curious events.

"Madlock won the championship. I didn't lose it," Griffey said ultimately.

"I can't think back about the batting title," Kenny said in the spring of '77. "What's done was done. Maybe I'll have another chance. People ask me about my potential and I really can't say. I feel that I have not reached the peak of my game, so who knows? My feeling now with the three-year contract is that I'm as much a part of this team as anyone else. I'm just turning 27 years old so there should be a lot of future."

A lot of future indeed. The long shot from Donora may be setting records after Morgan, Rose and Bench, the superstars who awed him, are in their waiting period for Hall of Fame consideration.

Phils' Larry Bowa a bit late

Mike Schmidt's frustrated dance

You're safe says Ed Sudol

Classic DP pivot over Dodgers' Davey Lopes

Angela, Joe, Lisa, Gloria

THE ULTIMATE

On that inevitable day when Joe Leonard Morgan joins baseball's immortals in the Hall of Fame in Cooperstown, New York, it will require all the oratorical and analytical skill of the introducer to summarize the abilities that add up to Morgan's credentials.

Babe Ruth and Lou Gehrig slugged their way into the Hall of Fame. Ty Cobb, Rogers Hornsby and George Sisler got there on a sequence of fat batting averages that in their better years exceeded .400. Pie Traynor has his bronzed niche in the Hall because he played third base as well as, and probably better than, any man who ever lived. Lou Brock probably will be there for having stolen more bases in a single season than anyone else.

Morgan is the epitome of the versatile player — the all-around athlete — the guy who can do anything to beat you.

Joe Morgan has won his campaign to be recognized as the most complete ball player of his time. The eventual induction into Cooperstown will be the culmination of years of dedicated mental and physical effort.

At 5-foot-7 and a mere 165 pounds at his top playing weight, the fact that Morgan is closing in on 200 career home runs going into 1977 makes him the all-time slugging king of the welterweights.

It is accurate to say that Morgan's size, or the lack of it, has provided him with the burning motivation to scale the heights that few baseball players have dared consider on such a broad scale. If there is one thing that Joe detests, it is the nickname of "Little Joe" and it is a name that Cincinnati media, fans and Morgan's teammates have learned to forget.

It probably never occurred to Bill Wight, a onetime lefthanded pitcher who was scouting for the Houston Astros in the San Francisco bay area in the early 1960's, that his success in signing Morgan was largely due to the fact that he never once mentioned Morgan's physical dimensions.

"If I heard it once, I heard it dozens of times from baseball scouts: 'you're a good little player but we can't offer you too much because your size might be a handicap'," Morgan says. "Houston's money offer wasn't as good as a couple of the other clubs, but Bill Wight is the only scout who never mentioned to me how small I was. He judged me on merit instead of size.

"All my life I've had to fight the idea I wasn't big enough to do things. Even as a little kid, I remember people saying 'he's a great LITTLE player.' I was the best hitter on my team in the amateur leagues and in high school, and I never thought anything about my size. I surely never thought it was a handicap."

If it ever was, Morgan has long ago turned it into an advantage in stoking his competitive fires. Even after winning the National League Most Valuable Player award two years running, Morgan feels enemy pitchers resent him hitting a home run off them because he looks like he should be a singles hitter.

"They don't mind me stealing a base or driving in a run with a single or double," Morgan says, smiling at the thought. "But I see it in their eyes when I'm circling the bases on a home run. They don't like a little man taking them downtown. I never say anything to them and I don't look at them. They do their job, I do mine."

The combination of assets that put Morgan in a virtual class by himself only start with his natural abilities. Joe has those, but he has a sharp, perceptive mind that never ceases to probe for ways to make more effective use of his skills. More than that, he has the determination to pay the price.

The Reds discovered Morgan's thoroughness when he joined Club President Bob Howsam, manager Sparky Anderson and other personnel on the club's winter press caravan a few weeks after he was acquired in the now-famous deal with the Astros on November 29, 1971.

"He told me he had a 'book' on every pitcher on our staff," Anderson said at the time. "Mainly it was based on the pitcher's habits and motion to help him steal bases, but he also detected some patterns that he thinks help him at the plate."

Cincinnati pitching coach Larry Shepard was skeptical when Anderson asked him to sit down with Morgan that spring and go over Joe's notes.

"I found out real quick that he knew what he was talking about," Shepard admitted. And he went further. He enlisted Morgan's help in lecturing the pitchers about holding runners on base.

That same spring, however, Morgan showed his own capacity to pick up tips. Ted Uhleander, a journeyman outfielder who had spent most of his time in the American League, was with the Reds.

"We were talking one day about pitchers," Joe explained, "and Ted mentioned that a righthanded pitcher had to move his right heel before making a throw to first base once he was planted on the rubber. Here's a guy (Uhleander) who doesn't steal bases but he had picked that up and I never had."

The Astros were starting their third season in the

majors when Morgan was signed after graduating from Castlemont High School in 1963 to play with their Modesto, Calif., team. Joe made a quick jump in mid-season to a higher level in the minors when he was sent to Durham, N.C., and was in big league uniform the last weeks of the 1963 season.

In his early days with the Astros, there was no hint of the monumental personality conflict that was to erupt between Joe and Harry Walker, who later managed the team. There were several positive influences on Joe's career, one of them being Nellie Fox, the longtime smallish second baseman who had his best years in the American League.

"Nellie Fox was a man I could relate to and he became a very special friend and instructor," Joe says. "If anyone helped me with my fielding, it was him. But more than that, he helped me recognize there was a place in this game for people of our size."

Billy Goodman, a high average hitter during his long career with the Boston Red Sox, served as a minor league batting instructor for the Astros and later as a coach. "Billy told me that no matter how old you get, you'd better come to spring training with the attitude you're a rookie who needs to win a job. If you've had a good year, prove it was no fluke. If you had a bad year, prove it WAS a fluke."

Morgan became the Astros regular second baseman in 1965, the year they moved into the then-fabulous Astrodome. He would celebrate his 22nd birthday on September 19 of that season, enroute to a satisfying .271 batting average. He also hit 14 home runs, raising some eyebrows early as to the power in his compact body.

Although Joe was born in Bonham, Texas, his family moved to Oakland when he was four years old and he considers himself a Californian. He enjoys a very close relationship with his father, who drove a city bus and later became a supervisor in the Oakland public transportation system.

"I wasn't born in a ghetto and never have lived in what you would call an all-black neighborhood. I guess that's the main reason I don't judge people on what they are and what color they happen to be. I remember once as a little kid coming home after having been in a fight and I told my mother it was 'with that white kid up the street.' She came down on me very hard, telling me 'You don't call people by their color' and that's the way it has always been with us.

"We've always been a close-knit family. There were six of us and I'm the youngest of the three boys. All of the family used to come out to see me play when they could and it meant a lot to me."

Morgan maintains the general impression that his years with the Astros were filled with turmoil is a myth. His problems with Walker were late in developing. The fact is Morgan had already played for two other managers in Houston before Harry took over in mid-season of 1968.

"I spent seven years in Houston and they were generally happy," he says. "I did not encounter off-field racial discrimination and my family and I were content and had no problems finding housing or other accommodations. I roomed with Jimmy Wynn most of those years with the team and we had a good relationship."

The Morgan-Wynn relationship was a contributing factor to Joe's difficulties with Walker in that Wynn ran afoul of Harry before Joe did. Walker has the reputation of being an astute batting instructor and had helped a number of hitters. But his concepts of hitting do not concur with those of some players and Morgan is one of them.

"I've always been a dead pull hitter," Joe puts it, "and that's definitely not Harry's style. He wants everybody to be able to hit to all fields, even big power hitters. I knew that I was more help to the team hitting the way I did. So Harry and I didn't see eye-to-eye on that.

"Harry had a clear cut philosophy. You were either with him or against him. When he and Wynn were having their problems, he assumed I was on Jim's side. But I didn't get involved directly. I have always spoken my mind and I think Harry resented the fact that I did a lot of talking with other players about what they were or weren't doing."

Walker, a man of strong Southern conservative leanings, had serious communication problems with players when he was at Pittsburgh prior to getting the Houston job.

In fairness to Harry, he probably was a victim of the generation gap.

Nonetheless, relations between the manager and Morgan were in a deteriorating state during the 1971 season, a factor that influenced the decision that sent Joe to Cincinnati. Morgan remembers one flareup vividly.

The Astros and New York Mets were playing on national television and Tom Seaver started for the Mets. Morgan had four straight hits in the game and was coming to bat in the ninth inning with the winning run at second base in a 4-4 game.

"Tug McGraw, the lefthander, was pitching for the Mets but I had ripped him for a single the time before," Joe recalls.

"Then I heard Harry yelling 'Joe' at me from the dugout. I thought he wanted to tell me something and started back and he said, 'I'm sending Wynn up to hit.' I didn't believe him. I screamed, 'You're what?' and all of a sudden I guess it hit him what he was doing to me. He said, 'Okay, you stay in and hit' but I was so mad I said, 'to hell with you' and stormed up the runway into the clubhouse.

"That's still the only time in my career a pinchhitter batted for me," Morgan went on, recalling the incident. "He threatened to fine me, but after a conference with (General Manager) Spec Richardson he didn't do anything. I guess Spec talked him out of it because I never heard anything more about it."

Walker and the Astros were convinced that they were in dire need of a power hitter for the cleanup role in their batting order and ex-manager Grady Hatton, the chief scout, had Cincinnati's Lee May near the top of the list. At the same time the Reds were out to shake up their lineup with added speed and defense. Negotiations between Richardson and Cincinnati's Bob Howsam began during the World Series and were still going on into the baseball meetings in Phoenix, Arizona, in late November.

"We honestly felt that in giving Tommy Helms, Lee May and Jimmy Stewart to Houston we might have helped them win the division championship," Howsam insists of what now stacks up as a proven one-sided transaction in his favor. "At the same time, we knew we were getting what we wanted and needed."

When Morgan moved into the Cincinnati picture, skipper Anderson told clubhouse manager Bernie Stowe to assign him locker space adjacent to Pete Rose — a move that paid exactly the dividend Anderson was seeking.

"When I came to the Reds I had seven years in the big leagues and, although I knew I had not achieved my full potential as a player, I thought I knew what it took to be a winner. But I found out differently. There was a price to be paid in discipline that Pete Rose, Tony Perez, Johnny

Bench and the others knew. I joined them at precisely the time they were rebounding from a bad year after winning in 1970.

"This was when the Reds were coming of age and I was fortunate enough to step in and even more fortunate to be aware of the need to pay that price."

Always a deliberate student of the game, Morgan's tendency to reduce baseball to an absolute and precise science took on new meaning in Cincinnati. For one thing, his new manager not only respected him for that approach, but encouraged it.

Meanwhile, it took Joe only a few weeks to become deeply involved in the seemingly acrimonious flow of insults that really is the glue that has held the diverse "star" personalities of the Reds together. This sometimes profane invective would make a stranger think he was on the scene of an impending gang fight. But with the Reds, it has been a way of life — and no player can be completely exempt. Morgan's sharp mind and natural tendency toward being a bit talkative eventually made him the ring-leader in the art of needling.

"Get away from here, Davey," he would say to shortstop Concepcion. "We're talking about hitting and you ain't no hitter." That specific line of reasoning served a serious purpose in that it increased the youthful Venezuelan's desire to become a better hitter.

"Seaver is pitching today, Bench" Morgan would say. "Aren't you sick or something?" The hard-throwing righthander of the New York Mets has a history of frustrating the All-Star catcher with hard sliders down and away.

The statistics of Morgan's first season in Cincinnati uniform would not be eye-popping. But even a casual glance at them reveals some indication of his value. His

.292 batting average was the highest of his career but he led the league in runs (122) and bases on balls (115) and had a career high of 73 RBI, a very substantial figure for a man who batted in the No. 2 spot in the batting order.

"Morgan turned the club around on the field as well as changed its image," Johnny Bench said in paying tribute to a man he beat out for MVP honors. "He was more deserving of the award than I was. The reputation of the Big Red Machine as a bunch of sluggers is done away with. The Machine has speed and plays a strong defense too."

The loss to Oakland in the World Series was a disappointment to Morgan in that he was appearing in his home town before a sizeable delegation of relatives and friends.

On the Sunday of Game Two, Jackie Robinson was a guest of Commissioner Bowie Kuhn in Riverfront. Hall of Famer Robinson had no connection with either the Reds or the A's but he was dying and baseball was quietly paying tribute. Almost totally blind with a diabetic condition, Robinson was escorted through both clubhouses by Joe Reichler, the commisioner's assistant and former veteran baseball writer for the Associated Press. As Robinson and Reichler stood somewhat awkwardly about to depart the Reds' area, Morgan quietly walked up to the man who had broken baseball's color line, shook his hand a second time and gently said, "Thanks."

His taste of a championship with the Reds stoked Joe's personal goals to new heights. Although he was an eight-year veteran, he was only 29 and the concept of becoming the finest all-around player in the game now seemed within reach.

"Most ball players reach a point where they want to concentrate on their strengths and in effect write off what they don't do well," Joe explains. "To Joe Morgan, the concept was develop everything to the maximum.'' Morgan has a tendency to use the third person approach to himself, hence reporters often hear him say, ''Joe Morgan isn't interested in leading the league in any given statistic.''

Morgan, who had been slowly accumulating college credits in the off-seasons, decided in the winter of 1972-73 to change directions in his belated educational pursuit. He switched from business to education and set a goal of becoming a college coach — a far cry from his $400,000-plus annual salary negotiated in the frenzied winter of 1976-77 when agents and lawyers buried the long-standing reserve clause that helped keep a lid on baseball earnings.

It must be noted, however, that as of this writing Morgan was inching ever closer to his goal of a degree.

Morgan's perceptive approach to everything is exemplified in an answer as to whether 1972 was his happiest season — the question being put to him late in that season.

''No, my happiest year was my rookie season in the big leagues with the sense of knowing I was a big leaguer. But 1972 is by far and away my most satisfying season in terms of accomplishment.''

His philosophy on stealing bases reflects the same in-depth analysis of his talent.

''With the Reds, I feel I should be selective in stealing bases,'' he explains. ''When everyone in the lineup is hitting well behind me, there's less need. But if Johnny (Bench) or Tony (Perez) is struggling a little, then I'm more conscious of the need. There also is the need in a close game in the late innings, or in the early innings of a game when we're up against a pitcher likely to give us trouble.

''As far as stealing third base is concerned, there

really isn't that much need to do it. You're in position to score from second on a single. To me, stealing third is risky, not because it's any more difficult than stealing second, but because the percentage is against it paying off in a run. That's what counts."

Despite his speed and sharp mental approach to stealing bases, Anderson found Joe reluctant to be a leadoff man.

"I've always felt I did better in the No. 2 spot in the order," he said at the time. "Your No. 2 man should be a guy who doesn't hit into many double plays, seldom strikes out and has good bat control. When I'm hitting second with Pete on first base, I get a nice hole on the right side to hit through. I like that. And even if I force him at second, I'm on base in a position to steal."

As Morgan's performance and reputation increased, he has made a science of tormenting pitchers with his patented long lead. In Riverfront he has both feet planted on the artificial green turf beyond the dirt-sliding area.

"I want to disrupt their concentration," he says of the pitchers. "The more they're concerned with Joe Morgan, the less concentration they have on the man at the plate."

Consistency on the field has been Morgan's trademark as he concentrated on improving the things he didn't do too well. He has for three years utilized arm--stretching exercises that are designed for pitchers to strengthen his own throwing arm. This has been the least impressive physical asset Joe has had to work with. His relay on the double play throw generally is regarded as ordinary and his long throws on balls going away from him have a tendency to 'die' a bit.

Joe keeps his own book on enemy pitchers and notes not only what they throw to him in given situations but what they are likely to throw to Rose, Bench, George Foster and his other teammates.

"I like to think I can read a pitcher's mind," he says. "Of course, I can't out-think him, but it helps to feel you can THINK along with him."

New pitchers in the league get as thorough an inspection by Morgan as they do from the scouts sitting in the stands making notes on them. "The first thing I go for is his movement to first base," Morgan explains. "The other things come next."

Morgan's candidness has made him one of the more interesting interviews in professional sports. Occasionally, his candor has gotten him into situations which he knows will draw criticism, but he doesn't mind.

"People seemed surprised when I said the Reds didn't show much loyalty in trading Tony Perez," Joe said in spring training of 1977. "But I was asked about the trade and I give honest answers. I don't want to make a big deal out of it, but when I was asked, that's the way I felt. Baseball is a business."

Morgan's toughest year physically was in 1974, when a sprained ankle, bruised hand and sore shoulder combined to force him to finish the season in considerable pain. It also was the year the Reds never caught up with the Dodgers in the National League West. Coming off the disappointing defeat at the hands of the New York Mets in the championship series the year before, the Reds put a lot of pressure on themselves that season.

In early September, Morgan reinjured his ankle in a game against the Dodgers in Riverfront. The next time at bat, with two runners on base, he swung and missed at a pitch and crumpled to the ground, his ankle giving out on him. Amazingly, he hit the next pitch out of the park to win the game and hobbled around the bases like a "Thank

God it's Friday" celebrant coming out of his favorite watering hole.

The championship season of 1975 was the year Morgan finally put everything together to his own satisfaction. Dropped to the No. 3 spot in the batting order to permit speedy young Ken Griffey to move in behind Rose, Joe concentrated on driving in 100 runs. He missed out only because he was rested with a back ailment in the late going after the Reds clinched the division title as early as September 7 over the outclassed Dodgers.

For the first time ever, Morgan had a batting average in excess of .300. His .327 figure had him in contention for the batting crown through mid-August. He drove in 94 runs, stole 67 bases and walked 132 times, the latter figure a Reds' club record.

Winning the MVP in a landslide vote vindicated a number of baseball writers in Morgan's judgment. Back in 1973, his name had been left off the ballot of 10 of the 24 designated National League writers, who had to list 10 players on their ballot. Moreover, there had been grumbling that black players were being overlooked generally in the MVP voting in that for six straight years white athletes had won it. Bench in 1970 and 1972 and Rose in 1973 accounted for half of those winners.

"This is a big day for me," Morgan said the day the vote was announced. "The margin by which I won is meaningful (he got 21½ of the 24 first place votes and teammate Rose the other 2½ with the split ballot being that of Bob Hertzel, one of the two Cincinnati writers on the panel), but most of all, the fact WE won the World Series makes it doubly meaningful."

Even then, Morgan emphasized that he had no intention of easing the rigorous demands he puts on himself. "This tops my season, but not my career. There are still things I want to do."

The Reds were a highly emotional team in the aftermath of their long-awaited World Series triumph and Morgan was very much involved.

"I was so happy when Carl Yastrzemski hit that fly ball to Cesar (Geronimo) for the last out in the Series that it frightened me," Joe said later. "It was the first time in my life as an adult that I don't think I was in control of my emotions.

"We're such a great team. The Reds ARE different. We're a unit and that's what separates us from the other clubs. Winning the World Series meant so much to me personally, but I was just as happy for the other guys. It has to be a different kind of happiness."

Despite his candor and a certain enjoyment of his own celebrity status, Morgan guards the privacy of his family and his life at home well. Gloria Morgan, his wife of 10 years, and their two daughters aren't seen on television or in the type of newspaper features one often sees on the families of prominent athletes.

"Gloria understands why it is important to us to have a private life," Joe puts it. "That's why she isn't that close to baseball. We've comc to an understanding on that."

Joe is aware that baseball travels and demands on his time take away from his role as a father, which is why he has no interest in staying in professional baseball once his playing days are over.

"There will be better things to do, and I want to be around to watch my girls finish growing up," Joe insists.

Morgan is also involved in a number of philanthropic and charitable activities, many of them black-oriented. He flies to Pittsburgh every winter to participate in a bowling tournament for Pirate slugger Willie Stargell's Sickle Cell Anemia campaign.

He is National Sports chairman for Dayton-based AIM (Adventures in Movement) for the Handicapped, an organization that utilizes dance and body movement to motivate deaf, blind, retarded and crippled persons, mostly youngsters.

"Joe has plunged into work with our group in a way that is a sheer delight," says Jo Geiger, executive director of AIM. "He's not along for the ride or just to let us use his name. He's a warm, interested person."

The demands on Morgan's off-season time force him to be very selective in the things he does.

"I've always been interested in other people," he says. "My mother and father and Baptist background have been an influence in that. I think of Nellie Fox, who did so much for me when I needed it. When he was dying of cancer, I called him frequently. I talked with him the day before he died and he's telling me how happy he was for the kind of a season I was having. That tells you a lot about him as a person, doesn't it?"

Going into the 1976 season, Morgan expressed some concern about the future, indicating he has not completely forgotten his small physical stature.

"I'm going to be more susceptible to injuries," he said. "I run risks not only when I'm running the bases, and that's every time I reach base because I'm forcing the pitcher to throw over and forcing myself to dive back, but in the field too. I hope I'm fortunate enough to escape any really serious injuries and I'll live with the others.

"Injuries, however, are the only things that can stand in the way of Joe Morgan's efficiency curve continuing to go up," he promised.

And go up it did as the Reds repeated as World Champions and Morgan repeated as MVP. He stroked a career high of 27 home runs and drove in 111 runs. His 60 stolen bases raised his career total to 505, putting him fifth on the NL's all-time list.

Like so many other Reds, Morgan credits batting coach Ted Kluszewski with helping him attain consistency. "Ted works with me whenever I need it. He emphasizes contact with good bat speed for me. It's not a new theory, but everyone needs reminding, I guess."

As he looks ahead with his contract making him the highest paid player the Reds ever have had, and one of the highest in the game, Morgan sees the top of the mountain in sight.

"If I didn't think I could keep improving to some degree, I'd have my retirement date in mind," he says. "Money will not keep Joe Morgan around this game if he isn't up to the standards he sets for himself."

Morgan disclaims any ambition to stay in big league baseball as a manager or a coach. "I don't want the hassles that go with managing," he maintains. "It isn't that I doubt my ability, it's just that I've had to pay the price to become the player I am. I don't want to pay it as a manager where the satisfaction would be less.

"I'm convinced that I could get a chance, although it still isn't easy for blacks to get consideration. Frank Robinson's break-through didn't open the doors very wide. There are plenty of qualified blacks who aren't getting consideration for these jobs. Maybe that will change, but it won't be because I'm interested in a job. When I leave baseball, it will be to settle down and enjoy a different kind of life."

Bench, Sparky debate Tom Gorman

The Commander strikes a pose

Now see here,
Mr. Frank Pulli

Shep, Alex, Sparky and George

THE COMMANDER

The taxi driver knew something was wrong. He had driven the man and woman, their five kids and their battered suitcases to the address the man had given him when he picked them up at busy Union Station. The adults seemed confused and it was after dark on a residential street. The kids were cranky and tired. So the anonymous driver didn't pull away immediately on this night in 1942 in Los Angeles.

LeRoy and Shirley Anderson felt very much alone in this big city, making the biggest and boldest move of their lives. It was a long way from the familiar world of Bridgewater, South Dakota, the family's starting point.

"The address LeRoy had given the driver was 35th Street," explained Shirley Anderson during the 1970 World Series in which her son managed the Cincinnati Reds. "Our relatives had rented this house for us, but it was on 35th Place. That driver helped us find it."

So it was that young George Anderson, not yet known as Sparky, became a Californian. That 35th Place address was quite close to the campus of the University of Southern California and its baseball field where Sparky Anderson's lifelong love affair with baseball began.

Rod Dedeaux, one of the most successful college baseball coaches of all time, noticed the little kid who stopped by every evening on his way home from school to watch his team work out. So Sparky became the USC bat boy and held that job to his great satisfaction until he was old enough to play high school ball himself.

Much gentle fun has been poked at Sparky's grammar. It happens to be improving due to the frequency with which he makes speeches these days — but it is still apparent that the man did not concentrate much on his formal education. Nonetheless, the University of Southern California intends to honor its onetime bat boy, and it is even possible that Dr. John Hubbard, the athletic-minded president of USC, will see that Sparky gets an honorary degree.

The odyssey of the Anderson family in pulling up its roots in bleak South Dakota and moving to California, where LeRoy Anderson did achieve a modest economic stability, attests to the humble roots of Sparky's life. LeRoy had been a house painter in Bridgewater. But that meant a lot of odd jobs and he was barely scratching out a living. The advent of World War II opened up the job market in big cities and the Andersons debated for several months about taking the plunge with their growing family.

"We finally decided to go," Shirley Anderson recalled. "So we shipped what furniture we had that was worth shipping and took off on the long train ride.

"We had to change trains in Omaha and there was an escalator in the station. None of us had ever seen one. George was fascinated by it."

After his Reds had swept the Pittsburgh Pirates in the 1970 Championship series, Sparky called his mother.

"Mom, go out and buy Dad a new suit and get ready to take an airplane ride up here for the World Series," he shouted. It was to be their first plane ride.

The reference to the new suit had a bittersweet meaning to Shirley at the time. "We had bought George only one suit in his life. Both he and his brother Bill wore it to their high school graduations."

Bill Anderson became a school teacher and, in Sparky's words, was "the one who made the grades in school for both of us."

But if George Lee Anderson was developing little taste for English literature or algebra, he was developing as a person with a deep-rooted addiction to honesty and friendliness.

"Sparky is too honest to be deceptive," said Richie Ashburn, many years later, about the man who was his teammate in 1959 with the Philadelphia Phillies. Ashburn, who made a successful conversion from the playing field to the broadcasting booth, was talking about Sparky on the eve of the 1975 playoffs.

"He's as good a manager as you'll find," Richie said, "because he knows his limits. Because he's so honest everybody on the team knows where he stands, and that's so important on a team with as many stars as the Reds have."

As his nickname suggests, hustle was Sparky's trademark as a player. A New York writer once jokingly suggested that Sparky must have been a Pete Rose type. "Are you kidding," he retorted. "I couldn't carry his shoes."

But Anderson was a good glove man as a player. His problem was an inability to hit the curve ball, but he had a fairly quick rise in the Dodger farm system after signing a contract for a tryout with the Santa Barbara club. The Dodger organization into which Sparky signed in 1953 was not the Los Angeles Dodgers of his home town. They were still Brooklyn-based.

Even though Santa Barbara was only a hop, skip and a few gallons of gasoline up Highway 101 from his home, Sparky was a lonesome kid when he reported to a manager named George Scherger, then only 33 but a man with six years of managerial experience in the low minors.

"George had a way with kids," Sparky said of the man he later hired to his big league staff. "He made you feel important, but not cocky. He was firm and I've never known a man who knows baseball any better."

Despite his mediocre season when he did get a crack at the big leagues, he was considered a good prospect and rose rapidly in the minor league structure of the Dodgers as a second baseman. He had trained with the big team in 1957 and 1958 and the Phillies gave the Dodgers a trio of players to acquire his contract at the winter meetings.

"The Dodgers were right," Sparky says. "The Phillies found out I couldn't hit the curve ball and, to tell you the truth, I didn't do so well with the fast ball either."

He actually hit .212 for a last place team and can never say he didn't get a good look. He was in 152 games. A demotion to Toronto in the International League wasn't hard to take and Sparky found a home at second base in the Canadian city, playing there four seasons before getting the manager's job for 1964.

As odd as it seems to those who know him today, Sparky had a trigger temper as a player. After his year with the Phillies, when he knew he was finished as far as making the grade on the field, he concentrated on making himself a manager. Conquering his temper became a lesson in self-discipline.

Jack Kent Cooke, the millionaire who owns the Los Angeles basketball Lakers and hockey Kings, owned the Toronto team during Sparky's first season as manager. Then he sold the team to the Boston Red Sox, who brought in Dick Williams for 1965 and fired Anderson.

All of a sudden at age 30, Sparky was in financial trouble. He had no baseball connection and a feeling of desperation was taking over. A man named Milt Blish, a baseball fan who lived in Thousand Oaks, the far north-west suburb of Los Angeles where Sparky lived then and still does, offered him a job selling cars.

"I was a terrible salesman," Sparky recalls. "I would never have lasted for anyone other than Milt Blish. He gave me every house deal (one in which there is no saleman's commission) that came along. He didn't realize it, but he turned my life around."

What Anderson learned then for the first time was a realization of how dependent he was on other people. Honest he had been. Now the compassion that marked his success as a manager was becoming an ingrained part of his being.

"When I was playing, I never cared about anybody," he says. Bear in mind, however, Richie Asburn's feelings about Anderson. Richie had been a star, a longtime fixture with the Phillies and his respect for Sparky had been established on pure friendship.

Sparky worked off and on for Blish in the off-season for several years. Through the summer of 1975, Milt was dying a painful death with spinal cancer. Sparky called his friend frequently, trying to be encouraging. After the Reds completed their sweep of the Pirates in the playoffs, a champagne-soaked Sparky said in his office, "I dedicated this championship to Milt Blish," much to the surprise of startled reporters from around the country.

Once he had his job sellings cars, Sparky wrote every organization in the majors asking about job openings — managing, scouting, anything. There was a call from St. Louis, from Sheldon (Chief) Bender, director of the Cardinal farm system who offered him a job managing the Rock Hill club in the Carolina league. Sparky jumped at it without asking the salary.

The boss in St. Louis was Bob Howsam, the man who ultimately hired Sparky to manage the Reds.

Sparky won pennants at Modesto, Calif., and St. Petersburg, Fla., in the Cardinal system the next two years. By 1968 Howsam had moved to the Cincinnati organization and Sparky was hired to pilot the Asheville Club, a Class A team in the Cincinnati system.

On the roster at Asheville was a young man named Bernardo Carbo, who was the No.1 pick in the first free agent draft in June of 1965. The Reds second choice in that draft was Johnny Lee Bench.

Bernie Carbo was and is an individual who marches to his own distinctive beat. A kid of Spanish descent out of Detroit, Mich., he was a moody youngster, given to doing odd-ball things. Anderson had first seen him in the winter instructional league the previous October.

Some of Carbo's teammates were calling him the "Village Idiot" and, rather than resisting it, he seemed intent on doing things to justify the nickname. Much to the chagrin of the Cincinnati organization, he was not concentrating on developing his apparent skills.

When Sparky took his team into Asheville to start the 1968 season, he called a meeting.

"I don't want to hear anybody on this team called Village Idiot or anything else as stupid as that," he said. "Anybody who makes fun of anyone else on this team without it being in a friendly or helpful way can count on not staying very long."

Carbo himself did not understand what his new manager was driving at, but over the course of the next few weeks Anderson lit a fire under the muscular youngster, who had hit a horrendous .201 the year before. Under Sparky, Carbo hit .281 with 20 home runs and within two years was in the big leagues to stay.

Expansion brought Sparky back to the big leagues. The San Diego Padres named Preston Gomez, who had risen through the Dodger organization to be a coach under Walter Alston, as manager. The Cuban gentleman, and no man in baseball can be called a gentleman with more accuracy, hired Sparky to be his third base coach.

At the end of the season, an even closer friend from Dodger days, wanted him to join the California Angels. Harold (Lefty) Phillips is the scout who signed Sparky to his first minor league contract "for $3,000 including salary" back in 1953. Phillips had been named manager of the Angels in mid-season of 1969.

You won't find the name of Lefty Phillips as a player in the major league records. Indeed, he was hardly ever more than a sandlot pitcher. He was a man who talked with the proverbial "mouth of mush," so indistinct that he almost seemed to have a speech impediment. But he had a keen baseball mind and Sparky Anderson loved him like a brother.

"Lefty Phillips taught me everything there was to know about playing this game," Sparky says. "He was a part of the Dodger organization when it was far and away the best in baseball."

On Oct. 8, Anderson picked up Phillips and they had an appointment to meet General Manager Dick Walsh for lunch, at which time Anderson would formally sign his contract to become a coach in the American League. On the drive down to Anaheim, Phillips mumbled that the Reds had fired Dave Bristol the day before.

After Sparky had signed his contract and after lunch with Walsh, who also came out of the Dodger organization, Phillips and Anderson were getting ready for the drive up the freeway to beat the rush-hour traffic. Walsh sent word from his office for them not to leave. He had taken a call from Bob Howsam who asked if Sparky could be released to become manager of the Reds.

When the news was given to Anderson, he could hardly believe what he was hearing. "Howsam wants you to call him when you get home," said Walsh, a man who had given Sparky a strong reference for that Rock Hill job a half dozen years earlier when the Cardinals asked for a recommendation. Phillips, nearly as elated as his friend, suggested they stop at his house for Sparky to make the call.

"They want me to fly there tonight," Anderson said, almost dazed. He was introduced at a press conference the next morning. "I flew in as George Anderson," he recalled. "I didn't need no phony name. Nobody knew me anyhow."

Jubilation was hardly the order of the day when the unknown Anderson was introduced as a successor to Dave Bristol, a manager popular with Cincinnati fans and even more so with his players.

"My reasons for firing Dave were that, in my judgment, he had developed as far as he could as a manager at that point for himself or our club," Howsam said in a 1976 interview. "At no time did I consider hiring a manager on the basis of his popularity to help sell tickets. I felt we had a ball club on the verge of great things, and I felt that Sparky Anderson was a man qualified to run the team."

Sparky himself was hardly overconfident in assuming command of a team with players the caliber of Pete Rose, Tony Perez, Johnny Bench and Bobby Tolan, even though he had put in one season in the farm system and had confidence in Howsam both as a man and as an executive. In retrospect, it seems amazing that Anderson surrounded himself with the caliber of coaching help he did.

Larry Shepard, who had managed the Pirates, was asked in as pitching coach and Alex Grammas, a man with obvious potential as a major league manager, was hired.

So was Ted Kluszewski, a longtime Cincinnati favorite. The one old friend Sparky added was Scherger, even more unknown than Sparky himself.

Anderson's grooming rules, which have become such a trademark, went into effect that very first spring. They reflect Howsam's conservative approach to life as well as to baseball. While Sparky concedes the code was strongly suggested by his boss, he had no difficulty adopting it as his own.

But it was the cooperation of Rose and Bench from the outset that made the rules stick.

"Pete came to me on our winter press caravan and told me he had been team captain under Bristol and would do whatever I wanted him to do as far as club rules went," Sparky recalled. "It was a great thing to do on his part."

Anderson has insisted always that the dress code which prohibits long hair, beards, mustaches and emphasizes 100 per cent conformity in wearing caps, socks and uniform have helped people accept the Reds for what they are.

"I honestly believe that part of our appeal to fans in the Cincinnati area is our clean image," he has said over and over. "When the Cincinnati Reds walk on the field, they have class. They don't look like trash."

If Anderson was able to convince Rose and Bench and the rest of the Reds to wear their hair short and accept his rules, he had a soul-searing struggle to get that point across to his oldest son, Lee.

Lee is really George Lee Anderson II, two years older than daughter, Shirlee, who in turn is two years older than Albert, the youngest Anderson heir. Lee was born in 1958, the year before his father's brief whirl as a big league player.

He was 15 years old in 1973 when he started to let his

hair grow long and eventually had it in a ponytail. Sparky told his son to get his hair cut before he returned from the next road trip.

"I didn't ask him. I told him," Anderson said in telling the story after it had an apparent happy ending.

"I came home and it wasn't cut," Sparky continued. "He was out in the garage working on his motorbike. I went out and we had words. I told him to get in the car, I was going to take him to get it cut. He ignored me. I realized the only thing I could do was fight him with my bare hands or back off. I didn't want to fight. So I just walked away. I cut him off from me.

"For over a year I had no communication with my son. He talked to his mother, but not to me. Then, before they had moved to Cincinnati in 1975, on our first trip into the Los Angeles area, he and I had a real, knockdown, dragout fight — not with fists but it came close to that.

"But out of that something seemed to give in both of us. We both learned something. Lee became a little understanding of what I'd been trying to say and I awoke to him, that I was wrong, handling it the way I did.

"Carol made me realize I hadn't given Lee that which he needed. I was thinking of my image and my image of his long hair. To me, it meant dope, rebellion, and it didn't have to. If you really love and communicate with your children nothing will come between you and them. But I let something come between Lee and me. I wasn't man enough to be the father my son wanted."

When the Reds flew back from Pittsburgh during the 1975 playoffs, Lee was in the front row at the airport. But his interest in baseball itself isn't all that great and Sparky has learned to let him march to his own drummer.

Albert, the youngest, often turns up in his father's conversation about "getting advice at home."

"This morning Albert asked me what I had to admit was a pretty good question," Sparky said once during the 1974 season. "He asked me if anything happened to Danny Driessen. And I said, 'No, and why do you ask?' He said he hadn't seen much of him and wondered if something had happened. Well, I guess I haven't been using Danny as much as I should."

On another occasion in Riverfront, Sparky had been roundly booed by fans who considered he was removing a starting pitcher too soon — this during the period his "Captain Hook" reputation was being built up. On their way home from the Stadium, Carol Anderson suggested that Shirlee tell her father why she was on her feet booing with the others.

"I didn't want anybody to know who I was," she responded, drawing a laugh from her father.

For all of his success, Sparky has managed to keep his family fairly well in the background. With all three youngsters beyond their early grades in school, they have remained in California until the end of the school term and then spent the vacation weeks in Cincinnati before returning to their West Coast home.

"One of these years, Carol is going to get to go to spring training with me and spend a month taking it easy," Anderson says, a rather unexpected comment from the manager of baseball's World Champions. But Carol Anderson's acceptance of her role as the mother of their children is hardly surprising in the light of their conservative life styles.

In his very first spring training, Anderson began to establish his reputation as an extremely cooperative manager with the media. One day after being filmed by two TV announcers and also taking time to do a radio tape, he jumped up in the dugout before a spring game and whispered to a reporter, "How come I'm so smart all of a sudden that everybody wants to know what I'm thinking about?"

The fast getaway of the Reds on the field in Sparky's rookie year as manager was a surprise to nearly everyone.

"We opened at home and then went on a western trip," Sparky recalls. "To be honest, I was worried about getting off to a bad start. I just never dreamed the club would do what it did."

What it did was beat Montreal in the Crosley Field opener, sweep a series in Dodger Stadium, then split four games in Candlestick Park to come home with a 6-2 mark. They went on to win 22 of their first 28 games to be six games ahead by May 7.

Wayne Simpson, the strapping youngster from Los Angeles who had pitched for Sparky at Asheville in 1968, stood the National League on its ear. The 6-foot-4, 210-pounder who swore he once threw a football the full 100-yard length of the field in high school, fired a two-hit shutout at the Dodgers in his major league debut and amassed a 13-1 record as the Reds made a joke of the race in the Western division. With southpaw Jim Merritt winning almost as consistently as Simpson, Anderson never had a moment of doubt in the regular season surge to the division title.

Simpson, however, tore a muscle in his right shoulder July 31. The Reds thought he would be back by the end of the season, but the sad fact is that the likeable kid never has made it all the way back, having bounced around in several organizations. Merritt hurt his elbow in September after winning his 20th game and, unfortunately, his career likewise went into an eclipse.

But the Reds were never in trouble in the Division race and put together a three-game championship series sweep over the Pirates to make Sparky a pennant winner as a rookie manager. Baltimore blitzed the Reds in the World Series, hardly surprising when one considers the Reds were so desperate for pitching that Merritt started one game with a left arm so saturated with narcotics that he could have been "busted" when he walked out to warmup.

"In looking back, I was an incredibly lucky man," Anderson said. "If 1971 had been my first year, or the things that happened to us in '71 had happened a year earlier, my career as a manager might not have been any longer than my career as a player."

In the off-season, Anderson was confident that both Simpson and Merritt would be very productive in his pitching rotation for his second year. But there was an ominous off-season development. Bobby Tolan, who had batted .316, played a fine centerfield and stolen 57 bases, ruptured an Achilles tendon playing basketball with the Reds team headed by Pete Rose. As it turned out, Tolan did not play a single game. Merritt wound up 1-11 and Simpson 4-7. Even more amazingly Johnny Bench, who had been the MVP the year before, went into an awesome tailspin, dropping from 148 to 61 in RBI and 45 to 27 home runs.

"It was a nightmare from start to finish," Anderson said, "but in the long run, it made a better manager out of me."

"There was never any thought given to changing managers," Howsam said in making the announcement of Sparky's third one-year contract on Aug. 31 of the lost year.

Sparky had been having numerous conversations with Howsam even before the formal rehiring about making changes in the club for the next year and the future in general. The basic goal was to emphasize speed and defense to take advantage of the more spacious Riverfront Stadium, which had been another factor in the demise of 1971 — the Reds found much of the power that resulted in extra base hits in cozy Crosley Field going to waste in Riverfront.

The end result was the fabulous trade with Houston that brought Joe Morgan, Cesar Geronimo, Jack Billingham, Denis Menke and Ed Armbrister in exchange for Lee May, Tommy Helms, and Jimmy Stewart, the versatile utility man. With Morgan adding a dimension of both on-and-off field strength with his personality and play, and a whopping comeback from Bench, the Reds climbed back to the top of the National League heap in '72 as Anderson's stature became accepted around the league.

"We're a much sounder, balanced team," Sparky could say. "We were a miracle team in 1970. If it was raining and we stepped outside then, the rain would stop."

There also was a major comeback in 1972 from Tolan — a player who eventually created one of Sparky's more difficult challenges.

A dramatic comeback led to a win over the Pirates in the championship series but the Oakland A's thwarted the ambition for the top prize in a tough, tense seven game World Series.

On the open date after his team had lost the first two games to the A's, Anderson discovered his wrist watch wasn't running. He walked into a discount store and picked out a $14.95 model of a popular-priced watch, much to the amazement of several clerks who recognized him.

"I got an expensive watch given me at a winter banquet,'' Anderson said to a reporter. ''I gave it away.''

The Series setback to Oakland was a bitter pill. In the spring of '73 Sparky confessed he shared a sentiment expressed by Rose and Bench — he wouldn't be happy now until he took a team all the way. He was pleased, however, to have been named National League Manager of the Year.

By now, the baseball world was beginning to respect Anderson for being a sound manager with the temperament to handle a team that included some of the biggest names in baseball.

''I'm a very lucky man,'' he countered, ''to be surrounded by the likes of Tony Perez, Joe Morgan, Johnny Bench and Pete Rose. They make my job easy. Nobody is going to walk onto this club and get out of line.''

The man who got out of line in 1973 was Tolan, who was unable to cope with the pressures of a miserable year on the field. The Reds eventually suspended him for insubordination, not to Anderson but to Chief Bender, the director of player personnel.

Tolan had been upset with the Cincinnati front office dating back to his 1971 injury and felt it unfair that he didn't get the kind of a raise he felt he deserved for his 1970 play.

He claimed what the Reds wound up giving him was a ''token'' raise in a contract he signed after the torn Achilles tendon.

Tolan and lefthanded pitcher Ross Grimsley were leading a verbal campaign against Sparky's rules against beards and long hair. Both were traded away shortly after the end of the 1973 season. A year later Sparky made an interesting observation about Tolan.

''My handling of Bobby Tolan was the biggest mistake I've made in management,'' he said. ''Somehow, I didn't see how disturbed he was and how much he was hurting inside. I missed the boat when I should have been trying to help him over the rough spots long before the trouble got out of hand. I feel in my heart somewhere I failed him."

Sparky Anderson, the compassionate human being, also regretted the fact that Bernie Carbo, after he was traded to the St. Louis Cardinals, complained that Sparky "lost confidence in me."

There was one man who never forgot one of Sparky's kind gestures. Pat Corrales, a journeyman catcher who was Bench's backup, was sent up as a pinch hitter and made the final out of the 1970 Series against the Orioles.

"I wanted to get my name in a World Series box score," Corrales said. "Sparky knew it, knew I might never have another chance. I'll always appreciate him thinking of me."

The 1973 campaign, in which the Reds charged from 11 games behind the Dodgers going into July to win the division going away, ended on a bitter note when the Reds lost the championship series to the New York Mets. This was the series punctuated by the battle between Rose and Bud Harrelson of the Mets, the bottle-throwing near riot and the stampede of people onto the playing field before the ninth inning of the final game was completed. Anderson sounded off against New York City and the people there — saying some things he publicly regretted.

Despite all of his early success, recognition of Anderson's managerial talents was slow in coming, especially from the general public. A lot of people dislike his "Captain Hook" attitude.(The nickname was given him by his own starting pitchers for his free-wheeling use of the bullpen.) Sparky still finds himself booed on occasion when he takes a trip to the mound to make a pitching change.

"I could care less about that," counters Anderson. "My pitchers know that and I think they understand it, even though I know at times some of them HAVE to get mad at me.

"We've had a great bullpen here from my first year on, starting with 'The Hawk' (Clay Carroll) and carrying through the guys we have now, and I've learned more and more to use 'em to prevent trouble as well as get us out of it. I'd rather bring a man in when he can afford the luxury of one mistake rather than when he can't make a mistake.

"It don't always work that way, but I'll look for any weakening in the sixth or seventh innings and move quickly. I tell our starters their job is to get us into the sixth and seventh.

"People complain about us not having complete game statistics. What do they mean? What counts is winning, and I figure that's what I've got guys out there for and I'm going to use them."

Anderson was booed early in the 1975 season for lifting John Vukovich, who started the season as the Reds' third baseman before the famous Rose shift, for a pinch hitter with the bases loaded in the second inning.

"My job isn't to win popularity prizes or build confidence," he said, defending his move. "My job is to win ball games. You ask yourself, the bases are loaded, a lot of times you play three or four games without ever having 'em loaded. So you ask yourself who do you want swinging the bat, Danny Driessen or John Vukovich. And you make your move and don't worry what people think."

Sparky went into the 1974 season with his second multi-year contract and a raise.

"I had never asked Mr. Howsam for more than a one-year contract," he said. "If it's good enough for Walter Alston, it's good enough for me."

The '74 season turned out to be the second in Sparky's five years that he did not win the Western Division. The Reds ran a close race all the way but never quite caught the Dodgers.

Anderson, confident, mature but still in his own way humble, was an impatient man through the 1975 regular season. The club had gotten off to a sluggish start when he decided to ask the veteran Rose to shift from the outfield to third base — a sacrifice that seemed astounding in view of Pete's age (34) and his established value elsewhere.

But with Rose at third and George Foster adding a power dimension to an already potent lineup, the Reds moved 3½ games in front by the 16th of June when Don Gullett, the youthful ace of the pitching staff, started against the Braves. The game was a 9-2 laugher for the Reds, until Gullett struck his pitching hand in the way of a line drive off the bat of Larvel Blanks and wound up with a fractured thumb.

As they had done a year earlier in a similar pitching shortage, the Reds called Tom Carroll, young right-hander, up from Indianapolis. Incredibly, they won 22 of their next 25 games and the Dodgers fell so far behind they were out of sight.

But there was no jubilation clinching the fourth division title under Anderson on September 7. "The champagne won't taste good until we've gone all the way," said the veteran Tony Perez, echoing Sparky's sentiments to a "T."

The Reds were all business when they faced the Pirates in the first game of the playoffs. Gullett, who returned to action in late August, pitched the winning distance and hit his first major league home run. Joe Morgan stole three bases, completely unnerving Pirate starter Jerry Reuss and the Reds won 8-3 to launch another three-game sweep over the frustrated Pirates.

The dramatic World Series that followed against the Boston Red Sox was hard on Sparky's nerves. Just as they had against Baltimore and Oakland, the Reds lost the opener, via a shutout at the hands of Luis Tiant, the ancient, likeable Cuban with the whirling dervish motion.

It took a ninth inning rally to pull out the equalizer in the second game, and then the Reds won two out of three in Riverfront before heading back to Boston for the kill.

Three days of rain pinned the Reds into their hotel headquarters. Howsam's assistants located an indoor gym at Tufts College available for a workout. When the driver of a chartered city bus couldn't find the place, the manager of the National League champions in full uniform walked into a service station to ask the astonished attendants for directions.

With play finally resumed, the Red Sox captured a 12-inning struggle that was saturated with drama as any World Series game ever played. The Reds had it wrapped up, seemingly, with a 6-3 lead in the eighth inning when Boston tied the game on a three-run homer off the bat of Bernie Carbo — the young man Sparky had rescued from the "Village Idiot" legacy nine years earlier.

It was uphill to triumph in the final game and at long last, Sparky Anderson had taken the team all the way.

Being atop the world hasn't changed Anderson's inner honesty and integrity. But it has changed him in other ways. There is an outer toughness that he can apply when needed, and he's made so many talks that his improved language and sentence structure would amaze his high school teachers.

Then came 1976. This time the Dodgers were a little harder to put away and there was another injury to Gullett — but once again the Machine was purring at deadly efficiency. The threat seemed to be from the Phillies, who had displaced the Pirates in the East and who roughed up the Reds in the early going of the regular season. But in late August, the defending champs beat the Phils three out of four in a tough series in Riverfront and when it came time to stare them down in the playoffs, they did it in three straight.

Whatever hard lines have developed in Sparky the manager, Sparky the man is still a concerned human being and his views of society are not overly rosy.

"We're in a sick society," Anderson says. "If you win, you're a hero. If you lose, you're trash. All we think about is the big money. Well, you take all the money you've accumulated in this world, and it won't let you live one more day.

"If the Reds win in 1977, people will cheer. If the Reds lose, they'll boo. But win or lose, I'll look in the mirror and see that's the same guy always looking back at me.

"If you think baseball is so important, go out to the children's hospital and then come out worrying because you ain't hitting .350!"

But the Reds have been winning, and it is to Anderson's credit regardless of what he says.

Back in the dark days of 1971, Bench started calling his manager "John McGraw" and Sparky confesses he loved it as a friendly gesture. But when the Reds swept the Yankees, winning the last two games in New York, Sparky became the first National League manager since the legendary McGraw to win back-to-back World Championships.

Not bad for a one-time Southern California bat boy, and a guy they called "Sparky Who" when he came to Cincinnati.

THE STAFF

One afternoon in Chicago's Wrigley Field in the early 1950's, a box seat patron was engaging noisily in the ancient practice of heckling the enemy. As so many are inclined to do, this gentleman went a step beyond the bounds of good taste.

His target was Ted Kluszewski, then the mass--muscled first baseman of the Reds whose playing weight was listed at a mere 240 pounds. From the safety of his seat, the man was referring to Kluszewski, among other things, as a "big baboon." The heckler felt a gentle tap on his shoulder and turned around to see a towering hulk of a man looking down at him.

"When you call that big guy out there a baboon," he said solemnly, "I have to feel you're calling me a baboon too. He's my brother."

The message from "Big John," the largest of the Kluszewski clan out of Argo, Illinois, turned the heckler into an ardent admirer. After all, "Big John" weighed in at around 280!

Ted Kluszewski is the "anchor man" of Sparky Anderson's coaching staff, which has enjoyed an almost uninterrupted tenure since it was formed in 1969.

As the new manager of the Reds, Anderson picked Klu as his batting coach; Larry Shepard, the then-recently fired manager of the Pittsburgh Pirates, as his pitching instructor; Alex Grammas, considered a good managerial prospect in his own right, as his infield tutor, and saved one spot for an old crony, the absolutely unheard of

George Scherger.

The only change came after the World Series triumph over Boston when Grammas was offered the job of managing the Milwaukee Brewers. His place on the staff was given to Russ Nixon, the onetime American League catcher and Tampa manager, as a reward for five seasons of patiently getting youngsters started up the ladder of the Red's farm system.

It is incredible that Kluszewski has turned into the astute hitting tutor that he is — "the absolute best in the business" in the words of Anderson. It isn't that Klu's own credentials aren't impressive. They are. But the big guy defied much of the logic in hitting and could not possibly have hit "by the book." He had no "wrist snap" whatever and propelled all his power from his arms.

In a man of lesser physical strength, Klu's style of hitting would produce mostly a string of gentle fly balls and routine grounders. But Ted's almost abnormal shoulder muscles provided the "whip" that his much-too-thick wrists couldn't.

Ted Kluszewski enrolled at Indiana University in 1944 on a football scholarship. A year later he played end on a team coached by the legendary Bo McMillin that brought Indiana one of its infrequent Big Ten championships. It so happened that the Reds trained on the Bloomington, Indiana, campus in the spring of 1943-44-45 because of World War II travel restrictions.

In the spring of 1945, several months before the Hoosiers won their football championship, Matty Schwab, the veteran groundskeeper of the Reds, spotted Big Ted taking batting practice in the IU Fieldhouse. He pointed him out to Deacon Bill McKechnie, the manager who had directed the Reds to pennants in 1939 and 1940.

"McKechnie wanted to sign him on the spot," recalled Warren C. Giles, then the president of the Reds. "If we had done that, there's no doubt he would have played for us in wartime 1945. We were an awfully bad team, but I knew we couldn't do that to our hosts when I found out how important he was in their football plans."

Giles, however, extracted a promise from Paul (Pooch) Harrell, then the Indiana Athletic Director, that if any other team attempted to sign Klu while he had eligibility remaining he would immediately notify the Reds.

In February of 1946 Ted married Eleanor Guckel, his high school sweetheart from back home in Argo. Although they are still a happily married couple, he learned early that two do not live as cheaply as one and he let it be known that he would listen to baseball offers.

Harrell called Giles late one afternoon in early May and warned him the Detroit Tigers were scheduled to meet with Ted the next day. Giles jumped into his car, took off for Bloomington and signed Ted to a contract before the stroke of midnight.

In the nine seasons he played with the Reds, Klu amassed 251 home runs, drove in 886 runs, fielded first base effortlessly but without exceptional range, and built a reputation as a likeable, genial giant who acted as a peacemaker when fights broke out.

On one occasion a fuss was breaking out between the Reds and Cardinals. Solly Hemus, an excitable St. Louis infielder, started toward the scene of combat snorting fire and fury. Suddenly he found himself lifted a foot off the ground with his feet still churning the air.

"Where you going, Solly?" inquired Kluszewski of the smaller man held firmly in his grip.

"Nowhere, Ted." he said, grinning.

"That's right, Solly, now just stay out of trouble," responded Big Klu who then stepped forward and pulled another St. Louis player away from the hostilities.

There was the time when the highly competitive Alvin Dark hit a twisting bouncer down the first base line which was fielded by Kluszewski a few feet in front of the bag. As Ted moved to tag him, Dark tried to do a fancy step around the big fellow. Wham — went Kluszewski's massive arm with the ball in his hand — into Dark's midsection and Alvin collapsed to the ground with a pair of broken ribs.

Klu departed the Reds as a player after the 1957 season, but never moved his residence away from Cincinnati. After his playing days ended, he went into the restaurant business and still has a part-interest in a downtown Cincinnati steak house that bears his name.

The Reds approached Ted in 1968 with a proposal to become a part-time minor league hitting instructor and he accepted with more eagerness than he cared to admit.

"I had worked just a little with Ted in the Florida Instructional League," said Skipper Anderson of his interest in Klu as his batting coach. "I liked the way he approached his job. He didn't take charge of a young man, he observed and made suggestions."

Which is how Kluszewski came back to the majors in his present role.

"All hitters obviously do not hit the same way and my job, as I see it, is to bring out the best in each player taking into account his style, his strength and some other characteristics," Kluszewski says of his low-key approach. "You look at a George Foster and a Cesar Geronimo and you know they have to be different type hitters."

Kluszewski and pitching coach Shepard share a dressing room-office, larger than Anderson's quarters in Riverfront, which was apparently designed for the whole coaching staff. It is also used, however, as the private "screening room" for both pitchers and hitters studying themselves on the video tapes the Reds take of all their home games.

Shepard is the dean of the Cincy staff (born April 4, 1919). He had no previous connection with Anderson before Sparky called him in October of 1969 to offer him the job as pitching coach. Shep had been fired a few days before the end of the '69 season, his second as manager of the Pittsburgh Pirates. A Pittsburgh newspaper ran a poignant picture of him walking across the Forbes Field infield, carrying a suitcase of personal belongings. The Pirates had finished third in the Eastern Division race with an 88-74 record.

"I was a very heartbroken and discouraged man at that point," Larry admitted. "I could see my baseball career behind me."

Shep has made his home in Lincoln, Nebraska, since his 1948 marriage to the former Joyce Hamilton. He had made some business contacts and investments there but he didn't think it over very long when the call came from Sparky.

"I did not want a collection of 'yes men' around me," Anderson says of his decision to surround himself with the very best available talent.

Larry had managed in the minors in the Pittsburgh organization from 1948 through 1966 when Gene Mauch hired him as pitching coach with the Phillies. It turned out to be a one-year assignment, similar to Anderson's one-year stay at San Diego, as an apprenticeship toward managing in the majors.

Shepard, a devout Catholic, is a man of considerable personal integrity who fits into the conservative pattern that Bob Howsam has established for his team. He has developed an excellent working relationship with his pitchers and stands up strongly for them in that he insists they have not gotten their full share of credit for the success of the club.

"Granted that we are a hitting team with the likes of Pete Rose, Joe Morgan, Johnny Bench and the rest," Shepard puts it, "but our pitchers have to be doing their share for us to have enjoyed the success we have. But somehow, they're the last to get recognition."

It is interesting to note that the Reds have won four pennants in the 1970's without the team ever having the one really big pitching leader of the caliber of a Tom Seaver, a Don Sutton, or a Catfish Hunter. Shepard's theories on pitching have some similarities to Kluszewski's concept of hitting in that they both stress basics.

"No matter how hard a pitcher can throw, he must eventually master a breaking pitch to survive," Shepard insists. He has been responsible for the emphasis on the "change-up," a pitch now taught at all levels of the Cincinnati farm system.

Shep is not without a sense of humor. One day in the 1976 season, the conversation in the dugout was about Mark Fidrych, the sensational rookie of the Detroit Tigers, who was getting headlines for his colorful style, including his habit of talking to the baseball.

"That's certainly nothing new," said Shepard. "I used to talk to the ball when I was pitching — but more than that, I used to smack it."

Jack Billingham had given up a couple of hits in a game in San Francisco a few days later when Shep walked to the mound. "Give me the ball," he said. Billingham did so and was startled to see Shepard smack it with his hand. "That's enough," Shep said to the pesky sphere, "I've killed you." A relaxed Billingham smiled and resumed pitching — a shutout.

Later, Larry related how he had "killed" many a baseball for his pitchers in the minors over the years.

"When a team makes a couple of errors or gets a couple of long hits, it's time to kill the baseball," he said. "It usually relaxed my pitcher. Of course, it's a trick you can't use too often."

Like most pitching instructors, Shepard concentrates on trying to get his men to pitch to spots. "You can't put the pitch where you want it every time," he concedes. "But if you're concentrating on location, you're thinking about what you should be doing and it's more often than not the concentration will bring the desired results.

"Pitchers are fragile and you have to treat them a little differently than you do other athletes. When a pitcher is having a rough time as a starter, he has four or five days to wait for his next chance to get straightened out. A four-game losing streak is three weeks long."

An arthritic hip condition came close to forcing Shepard's retirement, at least temporarily, in 1973. But he underwent off-season surgery that saw plastic and metal joints inserted into his leg and although he still walks with a stiffness, the pain is gone.

Shepard's seniority in age on Anderson's staff is a year and a half over George Scherger, the longtime "crony" Anderson rescued from obscurity in the minors.

Way back in 1940, at age 20, Scherger signed a contract with the Brooklyn Dodger farm system for $125 per month with a Class D club at Superior, Wisconsin. World War II interrupted an undistinguished playing career but the Dodgers gave Scherger a chance to manage in 1947 and he bounced around in the lowest level of the farm system through 1956. In a stop at Santa Barbara, California, in 1953 he had a 19-year-old infielder named George Anderson.

"George was the first person to influence me as to how much effort is required to win," Anderson said of his longtime friend when he hired him in 1970. "He always used to say, 'anybody can lose, it's easy' and that sounds awfully simple, but it's so important.

"George could be tough, but he was always fair and considerate. He knew kids like me could make a lot of errors and he tolerated that, but bonehead plays, mental mistakes and indifferent attitudes were something else."

Scherger gave up twice in his long managerial career in the low minors, but was lured back into baseball in 1967 to manage the Tampa Tarpons, which put him in the Reds' system. George had been spending his off-seasons working in an A & P grocery supermarket in Charlotte, North Carolina, work he continued through the winter of 1975-76, possibly making him the first wearer of a World Series ring to be stacking groceries.

"The hecklers in Tampa used to shout, 'unpack another case of tomatoes,' when I went out to change pitchers," George recalls. But he has a practical reason for working in the off-season even as a major league coach.

"We had four kids to put through college," he explained. "Our kids went to Catholic schools all the way and that cost us too. We've been catching up with the World Series money."

Scherger insists his anonymity is no cause for concern and he'll be eternally grateful to Anderson for giving him the opportunity to coach in the majors.

"When I was hired in 1970 I never dared dream I'd get the four years of service to qualify for the pension plan," he says. Going into the 1977 season, he's in his eighth year — not too bad for a guy who didn't break into the big leagues until he was 49 years old.

George took over as third base coach after Alex Grammas left to manage the Milwaukee Brewers. His instructional duties, in spring training and during the regular season, cover the outfielders.

"He's not too old to wield a mean fungo stick," quips George Foster of the coach who was 56 when the 1977 season began. "He can run you all over the outfield until your tongue hangs out." To the uninitiated, a fungo stick is a thin bat used by a coach to hit fly balls or infield grounders. The light bat serves the purpose as there is no impact of a thrown ball, the man just flips the baseball and hits it.

Alex Grammas, the fourth of Sparky's original staff, spent a two and a half season stretch as a Cincinnati infielder (1957-59) during a 10-season playing career. A native of Birmingham, Alabama, and a graduate of Mississippi State University (1949), the personable Grammas has been out of the big leagues only one year since coming up in 1954 for the first of two turns with the St. Louis Cardinals. In 1964 he managed Fort Worth in the Texas League for the Chicago Cubs but joined the Pirates in 1965 and spent five years as the third base coach there, the last two under Shepard.

"Alex was the man we wanted to coach third for us from the start," Anderson said. "He also knew the league better than any of us."

Grammas' father owned a candy company in Birmingham and Alex has been in partnership with Harry Walker, also a Birmingham native and a familiar big leaguer as both player and manager, in operating a supermarket in that Alabama city. One morning in spring training several years ago, Grammas sent a message to Walker with a writer going to the St. Louis Cardinal camp where Harry was a batting instructor.

"Tell him his half of the store was robbed last night," said Alex.

When the offer to manage in the American League came, Grammas said the most difficult part of the decision involved the feelings of his twin sons.

"They were just old enough (going on five) to start understanding baseball when we came to the Reds," Alex explained. "The Reds are the only baseball team they'd known and they knew all the players. It's going to be a tough adjustment for them." The Grammas twins, who have two older sisters, are Alex and Pete, sharing the names of their father, Alexander Peter Grammas.

The opportunity to manage the Brewers was too much to turn down, however, even though it meant leaving a World Championship team.

"I hate to leave, but I've been a coach for 11 seasons and at age 49 it's now or never," Alex said.

Anderson always joked about the slim build and dark hair of the even-tempered Grammas. "Here's a guy eight years older than I am who looks 10 years younger," laughed Sparky.

By coincidence, Russ Nixon, the new coach added in 1976, is almost as trim as he was as an American Legion player on the Cincinnati sandlots in the 1949-50 era. Slender he may be, but Nixon is a year younger than Sparky — who for the first time is older than one of his coaches.

Nixon's personality also matches Grammas', both being extremely pleasant, outgoing persons. Nixon had to earn his promotion by putting in five long years in the lower level of the Cincinati farm system at Tampa.

"I know what Russ had to feel going back there year after year, and he certainly was deserving of coming up with us," Anderson said. In Tampa, Nixon helped Dan Driessen, Ken Griffey, Doug Flynn and Will McEnaney get started up the ladder. Before that he had spent 13 years as an American League catcher with Cleveland, Boston and Minnesota.

"I was managing in the Mexican League when I got the call offering me the job," Russ said. "I'd do 12 hours a day hitting fungoes." That wasn't necessary with Russ taking over as first base coach and handling the infielders.

The Nolans: Gary and Carol with Kathy, Mark, Gary Jr. and Timothy

THE KID

They had put up a sign at end of town along California 20, the main highway in and out of Oroville (population 7,536) midway through the summer of 1967 proclaiming it the home town of Gary Nolan.

"That sign has gone down and been put back up more times than a yo-yo," says Gary Nolan, the likeable but quiet Cincinnati Reds pitching veteran with a whimsical impulse.

Chances are the sign has never really come down, but the gyrations of the pitching career of this tall, solemn-faced righthander have been enough to stir emotions wildly in Oroville. When the sign was erected, hometown hero Nolan was pitching and winning in the National League at age 19, having notched his first big league victory a year earlier.

Precocious is a mild term to describe Oroville's sometime No. 1 citizen. He married the girl next door, Carol Widener, while both were still in high school and was a father at the age of 17. If heredity and habit prevail and Gary, Jr. starts raising a family early in life, Gary, Sr. could achieve a somewhat mischievous ambition.

"I wonder what it would be like to be a grandfather and still be pitching in the big leagues," he muses. "Better yet, how about a father-son pitching combination?"

Nolan will be in his 12th season in the majors by his 30th birthday, although there are two distinct time gaps caused by the dramatic arm miseries that twice threatened to cut short his career.

Gary was the No. 1 selection of the Reds in the summer free agent draft of 1966, the year he graduated from high school. He signed for a bonus package in the neighborhood of $30,000 over and above his basic salary contracts. As a reward he was to be given a trip with the "big team" before being sent to Sioux Falls, South Dakota, to pitch in the rookie league.

With Tony Robello, the scout who discovered and helped sign him as his escort, Gary flew into Houston July 2 to join the Reds. Alas, the team was in desperate trouble. Manager Don Heffner never had gotten a firm grip on the team that had its morale shattered by the trade of Frank Robinson and other problems. The Reds dropped four straight in Houston and three more in Los Angeles before heading into San Francisco, only 140 miles from Oroville and where Gary had seen his first big league game.

Nolan and Robello did not get on the hotel-bound chartered bus at the San Francisco airport, and the quiet youngster's abrupt departure was the source of some gallows humor.

"I guess the kid has seen enough of us," quipped one player.

"I heard he's giving the bonus money back and wants to start all over," retorted another.

Actually, Gary was homesick for his wife and infant son and shortly thereafter took off for South Dakota. He compiled a fantastic record in a mere 12 games, blowing down 163 batters via the strikeout route in only 104 innings.

Invited to spring training in 1967 by Dave Bristol, the new manager, Nolan was an item of considerable curiousity although no one remotely thought he could stick with the big league team.

"I was waiting for him to pitch himself OFF the

club," Bristol said of his prodigy. "Instead, he kept throwing the ball over the plate and getting people out."

So Nolan beat the odds just as he said he would, came north with the Reds and won his debut over Houston with relief from Ted Abernathy in a 7-3 game.

In his second outing he came within one out of notching a shutout, this time against the Dodgers on the road. Alas, the Reds had scored only one run in the game and Jim Hickman's two out double in the last of the ninth tied the game, sending it into extra innings. Bristol then promptly lifted his young pitcher, who thus wasn't involved in the decision.

If Gary had been addicted to astrology, he would have asked the stars about the significance of that game.

All through his career he has been a pitcher for whom one of the strongest hitting teams in the majors has had difficulty scoring runs.

In his dazzling freshman year, Nolan pitched five shutouts on his 14-8 record and three of these were 1-0 victories, including one that went 10 innings. One of those classics was at the expense of the St. Louis Cardinals and flame-throwing Bob Gibson July 4th in Busch Stadium.

"I knew a lot of heat myself in those days," Nolan said, proud of the fast ball that was his pride and joy then. "In fact, I went to heat when I needed a strike or an out. My curve wasn't all that much."

Manager Bristol, who became manager of the Atlanta Braves in 1975, takes a different view of Gary's curve ball.

"If he hadn't had a decent curve ball, he'd never have made the big leagues as quickly as he did," Bristol insists, "I remember that game in L.A. After Hickman hit the double off a breaking pitch the next hitter was a guy who we usually curved. I was anxious to see what Gary would do. He pulled that cap down, stuck out his chin and struck him out on three breaking pitches."

The world seemed solid and bright for Gary and Carol Nolan going into the 1968 campaign. Their second son had been born and Gary had been given a significant raise from the minimum $8,500 for which he had pitched the year before. But on a very chilly day pitching against the Boston Red Sox in a spring game at Winter Haven, Nolan felt an ominous tightening in his shoulder.

The arm problem that followed seemingly verified the opinion of some baseball people that Nolan's jerky motion — one that made his fast ball more effective — could cause arm strain. Larry Jansen and Harvey Haddix, both of whom had pitched for the Reds at one time and were pitching coaches of the Giants and Mets respectively, both felt Nolan needed to alter his delivery.

"I was very frightened at what happened," Gary said later. "The doctors could find nothing seriously wrong, but the soreness was there. The arm didn't start to bother me until I got to about 75 per cent of my velocity."

Then the decision was made that Nolan would stay behind in Tampa, finish spring training with the minor league teams and pitch a few games for the Tampa team in the Florida State League.

It was a traumatic jolt for which Gary was totally unprepared. His two starts at Tampa were complete disasters. As a sharp pain shot up his arm, he dropped to his hands and knees on the mound in a blend of anger and frustration. Manager George Scherger, who two years later would turn up on Sparky Anderson's coaching staff in Cincinnati, called the home office to report that Gary was completely demoralized.

The Reds promptly summoned Nolan to Cincinnati where he was under the supervision of pitching coach Mel

Harder. His mental attitude and arm started to improve simultaneously and he was back on the active list in late May. He wound up with a 9-4 record that included seven straight wins and led the staff ERA a second year in a row.

The game he remembers best is a 5-0 shutout over the San Francisco Giants before a sizeable delegation of his friends from Oroville. The four-hit shutout should have been enough but he hit his first major league home run to cap a perfect evening.

"I'd been saving it for a special occasion," he deadpanned about the long ball.

By 1969 Haddix was the pitching coach for the Reds and spent much time that spring in what he said was "bringing Gary's arm and body into a more fluid motion."

The effort seemed about to pay spectacular dividends when he struck out 12 Dodgers in a 3-2 loss to Don Drysdale on Opening Day. But he pulled a muscle in his forearm pitching to a then not-quite-legendary Henry Aaron in his second outing. This injury was far more serious than the springtime problem of the year before. There were whispers that he lacked the physical and mental courage to overcome the fear of pain that all pitchers must endure from time to time.

His "shock probation" was an exile to the Indianapolis farm club for a brief period and then it was back to the Reds, active duty and a mediocre 8-8 mark. Sharp-eyed observers, however, noticed that the velocity of his fast ball wasn't quite what it once was — very tangible evidence that his arm problems were real enough.

"It's not much fun to think you're a candidate for Comeback of the Year at age 22," Gary said the next spring. "I think any pro athlete who has been injured or in the case of a pitcher who comes up with a sore arm, has an abnormal concern about it."

Meanwhile, his family now increased to three, Gary had purchased a home on a half-acre outside of Oroville and spent the winter working hard physically to get a good start for 1970.

That year was a turn-around for Nolan. Anderson's new coaching staff had taken over and Larry Shepard, Sparky's choice to handle the pitchers, spent a lot of time with Gary. While teaching him the change-up, which became an invaluable part of the repertoire of the "new" Nolan, Shepard took careful note to get a reading on Nolan's inner feelings. Sensing a concern about his future and a lack of confidence in the new manager and coach, Shepard and Anderson worked out a program to coax Gary out of the shell of his defensive thinking.

"I was worrying about my future and really had very little confidence at the start," Nolan admitted later. "But Sparky proved he had a very genuine concern for me as an individual.

"One night in May, I walked out to the mound against the Giants and for the first time in a year and a half I let it all hang out with everything I had. I ran out of gas and didn't finish the game, but that was a turning point."

With the Reds breezing to the Western Division championship, southpaw Jim Merritt becoming the club's first 20-game winner in five seasons and spectacular newcomer Wayne Simpson burning up the league, not much attention was paid to Nolan. But Gary picked up steam from June on, wound up with an 18-7 record and a nine-game winning streak.

"For the first time I was a pitcher with a scientific approach to the game," Gary puts it.

He pitched nine scoreless innings against the Pirates in the very first Championship Series game the Reds ever played but the old "bugaboo" of his mates not getting him any runs nearly did him in. They finally broke through for three in the 10th and Clay Carroll finished up the 3-0 win to start a sweep over the Bucs. Gary was the starter and loser of the World Series opener as Baltimore blitzed the Reds in five games.

The 1971 season was a disaster for the Reds and the only full-season losing campaign in Nolan's career. Even so, his 12-15 record was deceptive for he was mastering his change-up.

At long last, Gary was front-and-center as the Cincinnati pitching leader in 1972 and seemed headed for a fantastic season with things falling in place for the team. He was virtually unbeatable through June. His 13-2 record included a masterpiece against the hard-hitting Pittsburgh Pirates in which his change-up had some of the league's most dangerous lefthanded hitters chopping futilely away at the sharply-dipping pitch.

Then something popped in the shoulder, and in Chicago Gary advised Anderson he couldn't take his turn against the Cubs.

Nolan was an apparent cinch to be named to the National League All-Star team, an honor he wanted very much. Without considering the consequences as far as Gary was concerned, Anderson phoned Pittsburgh manager Danny Murtaugh, who was in charge of the All-Stars, and told him of Nolan's problem. Gary was deeply hurt that his manager didn't discuss the matter with him.

Although he was on the active list the rest of the season and started several more times, for all practical purposes he was out of commission. His 15-5 record and 1.99 ERA were great. Even with his shoulder problem, he started a game against the Pirates in the playoffs and the first game of the World Series against Oakland, giving up a pair of home runs to Gene Tenace as the A's won, 4-3.

The next season was the most miserable of Gary's life. He was on the active list most of the time but made only two brief appearances. Once more there were whispers he was mentally psyched out. Nolan insisted his shoulder hurt him. Team doctors developed several theories but nothing helped.

The nightmare persisted through the following spring when Nolan took matters in his own hands. He arranged for an examination by Dr. Robert Kerlan, an orthopedic surgeon in Los Angeles who had treated a number of athletes. The diagnosis was a calcium spur and surgery was performed in May. By late July he was throwing hard enough to join the Indianapolis farm club just to get back into a pitching routine.

"I had been advised that the odds were about 50-50 going into surgery that I would not have full freedom in the shoulder," he said later. "But those were good odds to me. Without some relief, I was finished. The hardest part of my ordeal was the realization that people, even some of my teammates, really didn't think there was anything the matter with me. They said it was all in my head. That was one of the reasons I moved out of Cincinnati and sold our home after the 1973 season. Back in Oroville, people didn't doubt my word."

Nolan's spectacular comeback in 1975 was instrumental in helping the team achieve its ultimate goal — the World Series triumph over the Red Sox. The Gary Nolan who won 15 games during the regular season had become a mature individual, who had inwardly turned to God as a practicing Christian and who had developed a sense of humor and a philosophy that reflected the inner turmoil he had survived.

"The doctors had told me my first season back would turn into a physical grind," Gary explained. "Things started off so well I was convinced it wasn't going to be that way. Then I did start to get tired, awfully tired, in the hot weather in July and August. But I was living with the power of positive thinking. When I needed help, I wasn't afraid to ask for it from the Man Upstairs. He's been good to me."

Although he had a start in the Championship Series and two starting calls in the World Series against the Red Sox, he wasn't involved in any decisions as the Reds savored the Series triumph.

"I was satisfied I had helped," he says. "Anytime you help keep your team in a game you're making a contribution."

Gary was the winner of the "Hutch Award," voted by major league broadcasters and players to a major leaguer who overcomes a major adversity. It is named in honor of the late Fred Hutchinson, the manager of the Reds from 1959 through 1964 when he died of cancer.

His third comeback influenced Nolan to bring his family back to Ohio and purchase a new residence.

There are those who feel his 1976 record, which was a 15-9 duplicate of a year earlier, was somewhat disappointing. But not to Nolan, who had the satisfaction of being the winning pitcher in the fourth and final game of the Series sweep over the Yankees.

"I won my 100th game during the season and notched my 1,000th strikeout," says Gary. "My lifetime ERA is still under 3.00. In many ways I'm happy merely to be around. If it's God's will that I win 20 games one of these years, I'll be delighted."

A lot of water has passed by Riverfront Stadium since Gary arrived on the scene. "The Kid" is pretty much grown up and those highway signs are still standing in Oroville.

Foster assaults helpless baseball

THE DIFFERENT ONE

If you were casting a black version of the musical " 'Lil Abner," George Foster would be ideal for the title role.

If you were producing a television series with a Black Superman, George Foster would be a cinch for the part.

If you were filming an authentic Tarzan movie, the role would call for an agile strong black man — George Foster again.

If you're beginning to suspect that this late-blooming slugger of the Cincinnati Reds is nearly too perfect to be considered authentic, you're on the logical road. But no matter how many contradictions you unravel in the muscular Californian, George passes the test of truth. It's unlikely truth at times; but he passes.

A non-drinking, non-smoking, Bible-reading bachelor, George's idea of a wild evening is taking a long time to decide which flavors he wants at the 31-flavors ice cream store. He also is a physical cultist, given to indulging in isometric exercises that helped build a rock-hard body. He has dabbled in philosophical and religious theories well beyond his formal education and he's a "loner" who nonetheless enjoys people.

The man who made a strong run in 1976 for that baseball rarity, the Triple Crown, can't put a finger on a day or week or even a year when his unique blend of thinking and attitude toward life began to take shape.

Foster's mother moved George and his sister from Tuscaloosa, Alabama, to Hawthorne, California, a suburb of Los Angeles adjacent to the area known as Watts. Thus George was a witness to much of the rioting there in the mid-1960's. He also came under the influence of a strong-willed grandmother, a staunch Baptist.

"I was taught the difference between right and wrong at a very early age and, naturally, I've always believed in God," says Foster, somewhat over-simplifying his philosophy.

As a growing boy, George was exposed to constant opportunities to head in the wrong direction, opportunities to which many of his youthful friends yielded. But he made baseball the anchor in his young life.

"I liked baseball so much that it helped me make whatever sacrifices were necessary for me to get through school," he explains. "I'd be inside doing homework when most everyone else was playing. I didn't go for partying because I wanted to stay in shape.

"The difference with me then, and it's still true today, is that I knew what I wanted. I could see my friends who were headed for big trouble. When you have a purpose in mind, it's not hard to pass up those other things. I was out there on the baseball field trying to find myself."

Even after his second World Series, Foster still spent the winter months in Hawthorne in the house he helped buy for his mother.

Not too many of his school friends are still around. A few of them have made good, but "an awful lot of them have been in trouble and a few are in prison," George says.

But Foster is no "goody goody" preacher-type wanting to theorize about the problems of others. He wants to help and has been active in visiting halfway houses and similar treatment centers for troubled youngsters in the Dayton and Cincinnati areas, as well as back home.

It would be a mistake, however, to assume that he has not had to weather some personal crisis points on his way to the top in baseball. And his success was painfully slow in coming.

Despite living in Dodger country as a boy, his idol was Willie Mays and his favorite team the San Francisco Giants. Because baseball scouts can look at talent the year around in California, a number of them help organize amateur teams.

The Giants got the inside track on Foster's services precisely through that sort of arrangement.

"This man, he worked for the Giants," explained George. "And he even called this team I played for the Giants. They asked me if I would sign if they drafted me, and I was taken in the secondary draft in January of 1968. I had enrolled in junior college by that time."

During the World Series of 1976, Tom Lasorda, just named manager of the Dodgers, was talking about the reports he turned into the office when George was playing in the Pacific Coast League.

"I wrote that he had a chance to become a very exceptional ball player," Lasorda said. "You could see the raw material, strength, throwing arm, speed — you name it."

It would be interesting to know why the Giants did not have the same strong opinion because George's rise in the farm system was quick. He hit a robust .321 at Fresno in his second season and was called up to the majors in September. That meant he trained with the Giants in 1970, but spent the season at Phoenix in the Pacific Coast League where Lasorda, managing at Spokane, Washington, saw him. He hit .308 there and was considered ready for the big leagues.

The 1971 season was the low point of the '70's for the

Reds, the distressing campaign starting after Bobby Tolan, the swift centerfielder, ruptured his Achilles tendon in a January basketball game. The Reds tried Hal McRae in center, then purchased Buddy Bradford from the White Sox. But when Tolan ruptured the same tendon again in early May, General Manager Bob Howsam was trying frantically to find another centerfielder.

At the time, Cincinnati had decided David Concepcion was their long-range man at shortstop, a position where they had a surplus of talent. Part of the surplus was Frank Duffy, a native of the San Francisco area and former Stanford University athlete. The Giants were interested in obtaining him and, meanwhile, Howsam had a nibble from the Pirates. San Francisco offered a choice of two young outfielders, Bernie Williams or Foster.

"Frankly, our reports on them were very close," Howsam explained. "I wanted to have a look myself and had Ray Shore join me out west." (Shore is the veteran scout who spends most of the summer evaluating talent on the other teams in both major leagues.) "The day we saw Foster, he did nothing impressive, but I had the feeling he had more to offer than the other players available for Duffy."

So on May 29, 1971, the Reds sent Duffy and pitcher Vern Geishert to the Giants for Foster — and George was far from delighted with the trade.

"I was with the team I had always wanted to be with and I was playing alongside my idol, Willie Mays. It really came as a shock when they told me I had been traded," he recalled.

Foster saw considerable service with the Reds, who were spinning their wheels through that troubled season. But there was nothing in his .240 average and his quiet ways to make fans or the media believe that the Reds had an undeveloped talent of George's ability in reserve.

It should be noted that Foster's soft, almost squeaky voice, coming out of such a strong physical body wasn't conducive to a casual estimate of his potential. He may be a Superman candidate today, but with his voice he could never sing "Old Man River" with any authenticity.

Tolan made a remarkable comeback in 1972 and so did the Reds, a development that shoved Foster deep into the dugout. His confidence was shot and he batted only .200 in brief appearances, although he did have one moment of glory.

The 1972 championship series with the Pirates found the Reds behind 3-2 as they came to bat in the ninth inning of the final game against relief ace Dave Giusti. Johnny Bench's dramatic home run tied the game and then Tony Perez drilled a single to center. Foster came in to run for "Doggie" and ultimately scored the run that meant the pennant on a wild pitch by the ill-fated Bob Moose.

When he was sent back to the Indianapolis farm club the following spring, the Cincinnati Enquirer headline read, with slight exaggeration, "Playoff Hero Sent To Minors."

That demotion turned out to be the crisis year during which George found himself. He was unprepared for Sparky Anderson's instructions to sharpen his baseball tools, even though the manager assured him that after doing that there would be a place for him in the Cincinnati outfield.

"I was not prepared for what happened to me, and it was a time when I had some doubts as to whether my goal in life was beyond me," he said. "Looking back, I know it was God's way of making me face up to some things."

But Foster had some help from mortals too. One of them was manager Vern Rapp, a clean-living disciplinarian, and another was Brooks Lawrence, the ex-pitcher who has influenced so many young Cincinnati players. Lawrence kept in close touch with Foster, offering counsel that George welcomed. And there was one final step — hypnotic treatment.

"I had read where Maury Wills had to overcome some type of fear and a hypnotist was a big influence in what he was able to do," Foster explained.

Wills had a deep-seated fear of reinjuring a leg that would deprive him of speed, his greatest weapon. Lawrence seconded Foster's decision to put himself at the disposal of a hypnotist. Among other things, Foster "discovered" that he too had a deep-seated fear of the baseball from having been hit in the head during his minor league career.

"I never even knew this or some other problems existed," George says. "I saw the man off and on for three months and it was a valuable experience."

Foster maintains his biggest gain out of his summer in Indianapolis was coming to grips anew with his relationship with God. "I was tested," he said, "I had to really believe that the Cincinnati organization, particularly Sparky Anderson, did have faith in me.

"That was the year I got the physical, mental and spiritual all together. Man is three-dimensional and he needs all of them to bring out his full potential. I had never had all three working before. My understanding and acceptance of God finally was in place. I had to re-discover myself."

Even so spiritually fortified, there still was the problem of Foster finding a place to play in a Cincinnati outfield where Pete Rose was king, Cesar Geronimo finding himself and Ken Griffey knocking at the door. On top of that, the club traded dissident pitcher Ross Grimsley, rebelling at the strict dress code of the club, to Baltimore for Merv Rettenmund.

Foster still hadn't made it all the way to the top through the '74 season, but he was able to cope with the frustration of appearing in only 106 games.

"The thing is, in my mind, I felt I was going to be traded," George said. But Anderson was opposed to that move because he sensed Foster was on the verge of exploding to his full potential.

The big fellow's breakthrough came in early 1975 as a result of what Sparky now calls "my biggest brainstorm." That was in asking Rose to move into third base to create an opening in the outfield for either Foster or a platoon arrangement involving Dan Driessen.

"In my mind, George was going to get the good crack at the job," Anderson insists. "I mentioned Danny at the time but George is a fine defensive outfielder. And then the baseball started to jump off his bat."

Despite his quiet nature, Foster has a bit of gentle showmanship in him. His use of a stained black hickory bat, instead of a conventional style, is a gesture, as is his one-handed grab of fly balls in the outfield.

"I can catch them that way and I think the people like it," George says with a smile. "I suspect if I had come up through the Cincinnati minor league system they might have made me change. If I ever drop one, they might even try it now. I can use two hands. In fact, I will if I have to make a strong throw."

Confidence is his one-word explanation of his batting heroics.

"I walk up to the plate feeling I can make contact if I'm ready," he says. "I respect the pitcher but in my heart I know he's no better than I am." Pitchers, however, annoyed with George's habit of stepping out of the box between pitches to concentrate, might question whether he respects them or not.

Cincinnati followers had no difficulty recognizing Foster's contribution to the cause in 1975 through the surge to the dramatic World Series conquest of the Red Sox. No member of the Machine, Johnny Bench or the departed Tony Perez, ever hit a longer home run than the one Foster hit into Riverfront Stadium's gold seats off Houston's Larry Dierker during that season.

In summation, Foster wound up with an even .300 batting average. At one stage he had a 16-game hitting streak and in the division-clinching game against his onetime team, the Giants, he went four-for-four. He continued to hit well against both the Pirates in the playoff and the Red Sox in the series.

Being a member of the World Champions and cashing a World Series check, not to mention acquiring the Series diamond ring, did not change George's life style to any degree. He still lives in a downtown Cincinnati hotel during the season as he has since coming to the Reds. His habits as a loner make that more acceptable.

"I have my tapes and listen to music when I need relaxation," he explains. "I read a lot. Life in the big leagues isn't one exciting day after another."

In 1974 a retired Detroit sports writer named Watson Spoelstra, who had become a "born again" Christian, started a movement called Baseball Chapel. It sponsors a non-denominational religious service each Sunday morning in the clubhouses of major league teams. Foster and Gary Nolan were charter members of the Cincinnati group where attendance usually runs about 80 per cent of the uniformed personnel.

Foster also is involved with Bible study classes at different times with Wendell Deyo, a former college football player who is in the paid ministry of the Fellowship of Christian Athletes in Cincinnati.

It is an additional bit of the Foster legend that when the Reds had their champagne-popping clubhouse celebration after capturing the 1976 Championship Series, George and teammate Joel Youngblood had sneaked into the recreation room and were playing ping pong.

"I didn't have anything against it," he said of the celebration. "I just didn't want to be part of it."

The confident, poised and mature Foster who took the field in '76 made a run for the Triple Crown (winning a league batting, home run and RBI championship the same season) which hasn't been done in the National League since Hall of Famer Joe (Ducky) Medwick did it in 1937. Although George didn't quite make it, he was voted National League Player of the Year by his fellow players and was MVP of the All-Star game for his three hits, including a home run, in a 7-1 conquest of the Americans.

"I started swinging for home runs late in the season and I wasn't getting any," Foster said. "So my batting average slipped. There's a lesson there. Don't be greedy. I'll remember it."

Marriage apparently was still far into the future as George headed into '77 with a three-year contract that elevated his salary into the six-figure plateau.

"I feel I belong," he says, "and if I keep my health my best days are ahead of me."

Indeed they seem to be just that for the youngster who knew right from wrong at an early age and hasn't forgotten those simple basics along the way.

Odd, isn't it, that such an attitude stamps him as different in this day and time?

Kid Gullett unloads a hummer

THE KENTUCKY KID

The secretary in the office of the San Francisco Giants had thrown the reservations clerk at United Air Lines a curve. She wanted to book Thomas C. Sheehan into a place called Lynn, Kentucky.

It so happens that Lynn isn't even on the highway maps of the Commonwealth of Kentucky. But after five or six telephone calls, the chief scout of the West Coast team finally had an itinerary. Then a robust 74-year-old Thomas Clancy Sheehan flew into the Greater Cincinnati airport, squeezed his 260-pound bulk into a rented car and began the remainder of his journey on a meandering two-lane highway along the Ohio River — "upriver" into Greenup County.

Chances are in that spring of 1969, Sheehan never had heard of Geenup County's most distinguished and best-known citizen — novelist Jesse Stuart.

But by October of 1976 Greenup County's best known citizen was Donald Edward Gullett. The hero of a pair of World Series efforts on behalf of the Cincinnati Reds, he had stirred up deep feelings by utilizing his right to become a free agent and sign for the gold and glitter offered by the faraway New York Yankeees.

At age 18, young Gullett was what is known as a "pheenom" in baseball parlance, a youngster of unbelievable ability. The previous fall the sturdy 190-pounder had scored an amazing 248 points as a running back in 13 football games for McKell High School. He had one game in which he scored 12 touchdowns, a 72-point outing against Wurtland, one of his school's biggest rivals.

"Our coach was mad at the other coach," Donald confided to reporters in his first big league training camp. "So he sort of turned me loose."

Gullett also averaged 25 points as a basketball guard but it wasn't the college recruiters who were scrambling to Lynn, the community that literally wasn't on the map.

On his drive up to Lynn, Tom Sheehan was doing some introspective thinking. Beating the bushes was finally getting to be a chore, but baseball had been his life. He never tired of telling stories about his years in the game, including the one about pitching in pain the last half of the season in the American Association in the 1920's on what later proved to be a fracture in his lower leg.

"I kept telling old Mike Kelly it hurt like hell," Sheehan would say. "He kept saying I'd be all right my next time out. Finally at the end of the year I had it X-rayed and it was broken."

When he got back to San Francisco, Sheehan turned in a strong report on Gullett but advised owner Horace Stoneham that he'd better find a younger scout for tracking down prospects in places like Lynn, Kentucky in the future.

Cliff Alexander, a scout for the Reds who lived in Hamilton, Ohio, and knew some better roads across southern Ohio to track down Gullett, saw him pitch one game in which nary a runner reached base. It was almost a perfect-perfect game. The Kid struck out 20 of the 21 batters he faced.

That spring Gullett struck out 120 batters in 52 innings. Alexander's report gave him a "Can't Miss" label and the Reds backed up that opinion with one from Rex Bowen, who spends most of his time scouting at the major league level.

Alexander's estimate was lavish enough to make a limited comparison with the great Los Angeles Dodger southpaw, Sandy Koufax. Cliff had more than a passing knowledge of Koufax as a pitcher and a person in that his sister is Mrs. Walter Alston, wife of the longtime, but now retired, manager of the Dodgers.

Cliff Alexander died of cancer in 1975 but he had lived long enough to know that his estimate of Gullett's potential was fully vindicated.

The Reds were delighted that 13 other teams did not have Gullett tabbed as their first choice in the nationwide draft, so the Ohioans quickly claimed him when their turn came around. Sent to faraway South Dakota in the rookie league where Gary Nolan had flashed so brilliantly three years earlier, southpaw Donald compiled a 7-2 record with a league-leading 1.92 ERA for Sioux Falls.

He pitched that October in the Florida Instructional League where Sparky Anderson, who had just been named manager of the Reds, had his first look at him. The following spring, Gullett pitched his way onto the major league club as Nolan had done three years earlier. But there was a difference in that Anderson and pitching coach Larry Shepard did not consider him in a starting role. Rather, they dared hope he could relieve — at the tender age of 19!

"He made the club because he threw nothing but strikes and we had no lefthander," Shepard said, looking back.

"We had Wayne Granger and "The Hawk" (Clay Carroll) to carry the load," Anderson said. "We took a chance on Don because we felt he could get one or two lefthanded batters out."

Gullett was strictly a one-pitch workman then — his fast ball with blistering velocity.

If anyone ever takes on the project of an "in depth" biography of the likeable Kentuckian, it will be a chore. There simply are few if any anecdotes, in print or in memory, of his early days.

"If you wanted to ask Donnie something, you had to go look for him," said Joe Nuxhall, the onetime pitcher who has been a mainstay on the Reds' radio broadcasting team for years. "He absolutely had nothing to say."

Anderson also feels that the fast breakaway of the Reds in that 1970 Western Division race put little pressure on either Gullett or the board of strategy.

"We had no problems, and it was easy to pick spots for him," Sparky explained. "That isn't to say he didn't do a job. He did — and later in the year we had more use for him."

Gullett's total work load in the regular season was limited to 78 innings, but he won five of his seven decisions. More important, he pitched well in the championship series with four scoreless innings against the Pirates. But even before then, the decision had been made to convert him into a starter.

The silent Kentuckian displayed a stubborn streak when he refused to go along with the club's suggestion he pitch winter ball in Puerto Rico to add a curve ball or pershaps a slider to his repertoire. But Donald was adamant about wanting to spend the winter at his home, having married blonde Cathy Holcomb the previous January. He didn't budge even when the Reds said he might have to spend the 1971 summer at Indianapolis if he failed to pass a "crash course" in spring training. So he went back home to Lynn and spent his free time tramping the hills, shooting birds "in season" and doing the things he'd always loved to do.

Shepard's so-called "crash course" wasn't all that

demanding. "We knew we were dealing with a very special arm and we were going to proceed slowly," the veteran teacher said.

Gullett, meanwhile, could be backed into a corner for an interview in a manner of speaking.

"It took me a while last season to convince myself I really belonged in the big leagues," he conceded. "I didn't really know what was going on. I just went out there and tried to throw strikes. But when I did well in the playoffs, I knew that I really did belong."

Anderson's plan for Don's in-season efforts called for careful attention to the number of pitches he would make. An arbitrary limit of 100 was prescribed.

Donald's conversion to starting was a fantastic success, a development not fully appreciated in the wake of the tumble taken by the 1970 pennant winners, who were never even a remote factor in the race the next time around.

"Don's control was an absolute revelation," Shepard said late that season. "He was making unbelievable progress."

The still-quiet kid from Lynn walked only 64 batters in 218 innings, posted a 16-6 record out of 31 starts and notched a 2.65 ERA. As Gullett celebrated his 21st birthday in January of 1972, established stardom and really big money seemed just around the corner. But the first of a series of serious physical problems that haunted him through the rest of his days in Cincinnati cropped up.

He came down with a fever, feeling quite below par for several weeks as doctors groped for an accurate diagnosis. He was still weak, his weight was off and he really was a concerned young man when he reported for spring training. Another diagnosis indicated he had contracted hepatitis and his strength would be slow in returning.

The result was the only losing season he's ever had — a 9-10 record, but even that was misleading. Over the last six weeks of the season, he was as effective as anyone on the Cincinnati staff which had been bulwarked by the off-season trade addition of tall Jack Billingham.

No athlete ever enjoyed a better or closer relationship with his manager or pitching coach than Gullett. Donald's quiet life style, his unselfish attitude and his strong competitive instinct appealed to the basically conservative instincts of both Anderson and Shepard.

"I'd be proud to call him my son," Shepard said during spring training of 1974. "He's such a sincere, honest, coachable person. All of that and a lot of God-given talent."

The occasion was a discussion of Gullett's 18-8 record in '73, a season which started off rather poorly for him. He was having spotty control troubles in his early starts and was given a temporary exile to the bullpen. That move proved a blessing in disguise because all of a sudden Gullett started to master the fork ball that Shepard was working on as an auxiliary weapon to go with his natural heat.

His curve and slider never had really developed but the fork ball was just what he needed.

"The great pitchers in this league always have been able to overpower you," Shepard emphasized. "Bob Gibson, Sandy Koufax, Tom Seaver, Steve Carlton — you name them — they all could blow you down. But they all reached a point where they did need a good second or third pitch."

Despite his four-week banishment to the bullpen, Don had an 18-8 season. Late in the year he was working on a no-hitter against the Chicago Cubs but yielded a one-out double in the eighth inning to Paul Popovich, a light-hitting infielder who usually made bat contact but without much authority. He finished the 1-0 game with a two-hitter and felt pretty good about it — until the next day.

"I guess you know that was a lousy pitch you made to Popovich," Shepard challenged him.

"I didn't think it was," retorted Gullett, who had tried to hit the outside corner with a low and away slider which Popvich, a righthanded hitter, blooped to right field.

"It was not a good pitch," thundered Shepard. "You had let him see it twice before. In that spot, you challenge a hitter like Popovich. You don't toy with him or tease him. You go get him. You're better than he is."

All the while, Gullett was fighting physical woes not always apparent or written about. He always has had a slightly tender left ankle, a souvenir from an old McKell High School football injury. He had the ankle taped every day he pitched. And late in the 1974 season the gritty Kentuckian came up with a back problem that restricted his efficiency.

"It still was my best year in terms of complete games and ERA," Gullett said. "Shep keeps telling me I can become a super pitcher. When you're a super pitcher, you might lose but you never get blown out of a ball game. I've had a couple of bad games, but I always felt I could work out of any tough spot I find myself in."

The talk in spring training of 1975 once again centered on Gullett attaining his full potential.

"He's about this far from greatness," said Shepard, spreading his thumb and first finger about two inches apart. "I know that we have at times expected too much from him. But he's got to become our take-charge guy."

"If I don't win 20 games this year, I'm going to be

disappointed,'' Gullett said, accepting the challenge.

It must be noted that Don now had an agent, the legendary or infamous Jerry Kapstein, depending upon your point of view. His 1975 contract, negotiated under the old reserve clause that Kapstein and others were destroying, called for remuneration of $85,000.

Once the season started, the southpaw flame-thrower seemed on target toward his 20 victories. His record was 8-3 when he took the mound against the Atlanta Braves in Riverfront the night of June 16. His ninth win was a routine one that was abruptly turned into a near-tragedy.

Gullett had a 9-1 lead as he went to the mound in the ninth inning and seemed to be running out of gas as three of the first four batters he faced rapped singles. Then shortstop Larvel Blanks, who goes by the unlikely nickname of Sugar Bear, stepped in. He ripped a line smash straight back at Don, who instinctively put his bare pitching hand down to intercept it. The ball struck his hand with full impact, he sank to his knees in pain and then got to his feet looking at his hand as if in a daze. The imprint of the ball's stitches were stamped across the base of his thumb and into the fleshy area of his hand.

''I think it's broken,'' he said to trainer Larry Starr and Anderson. Starr asked him to move his fingers. He couldn't and the trainer knew it was broken.

''The Kid'' did not pitch again until August 18 when he started against the St. Louis Cardinals, a team against whom he had enjoyed spectacular success. All he did was pitch five scoreless innings and then spent the rest of the season getting back into the groove for the post-season competition, finishing with a 15-4 record.

Gullett was the starter in the opening games against both the frustrated Pirates and ultimately the Red Sox in the World Series. He not only pitched a route-going 8-3 win over the Pirates, but hit his first professional home run in the process.

The start in Boston was not so successful. For six innings he and ageless Luis Tiant, the amiable Cuban with the whirling dervish motion, were locked in a scoreless duel. But the Red Sox exploded for six runs in the seventh and Tiant pitched a five-hit shutout for Boston's 6-0 win.

The Series was deadlocked at two games each when Gullett started again in Riverfront. This time he was a 6-2 winner with ninth inning help from Rawly Eastwick.

Under normal circumstance he might not have started again, but the three days of rain that delayed resumption of play after the Series shifted back to Boston set him up for the seventh game — if necessary. It turned out it was but Donald was hardly "vintage" Gullett. He walked in two runs as Boston scored three times in the second inning. Anderson stuck with him and he finally got the side out but he was long gone by the time the Reds rallied to win the Championship.

That moment turned out to be Donald's swan song as a contented member of the team.

The winter of 1975-76 was the most unsettled in baseball history as the Players Association and club management continued to debate the renewal of the "basic agreement" — the battle being over the reserve clause which, in effect, had been rendered meaningless by adverse court rulings. Spring training was officially delayed and the spectacle of ball players training on their own in T-shirts, sweatsuits and assorted other garb became commonplace in Florida.

But while most of the Reds came to the Tampa area and worked out on their own, Gullett was holed up in the expensive Spanish-style home he had built in Lynn while Kapstein tried to negotiate a multi-year contract on his behalf. The Reds, to that point in time, had never given anyone a multi-year contract although catcher Johnny Bench was about to sign a two-year pact.

Finally, Commissioner Bowie Kuhn's intervention formally opened the camps and the parties eventually came to terms on a renewal of the basic contract, but the old option "hammer" held by management was gone.

Gullett was the only Cincinnati player who went into the season unsigned and thus became a potential free agent for the first such draft scheduled for November of 1976.

People around the Kentuckian noted a change in his attitude. Once quiet by nature, he was now reticent almost as if under some spell. He refused to discuss the contract problem. More alarming to Anderson and Shepard was the fact he had a stiff shoulder and never was completely free of the problem all season. He managed to start only 20 games, working 126 innings, but it is still a tribute to his ability that he posted an 11-3 record and pitched the first games of the Championship and World Series once again.

This time he was a 6-2 winner over the Phillies and a 5-1 winner over the Yankees. But in the latter game he suffered a sprained ankle which was put in a cast and he accompanied the team to New York on crutches.

"There was a 'hardness' in Gullett that had never been there before," said a teammate. "He seemed suspicious of anyone and everyone in the front office, and intimated that he had never been properly compensated over the years."

In fairness to Gullett, as of this writing, he himself never had publicly made that complaint, even after going through the free agent process and signing a seven-year agreement with the Yankees worth an estimated $2.1 million.

"Looking back, I'd say we never really had a chance with Gullett," said Dick Wagner, then the executive Vice President and since promoted to General Manager. "It wasn't so much the money, which we would have been in a position to discuss, rather it was the length of the contract. With them, it was five years or nothing."

Friends of Gullett told writers that Don's hard line was pretty much motivated by the fact he was reminded in negotiations that the Reds won both the 1975 and 1976 division races without a contribution from him befitting his status as the team's best paid pitcher.

Judging by the 1977 salaries the Reds are paying Joe Morgan, Dave Concepcion, George Foster and the other stars, chances are Gullett could have come close on an annual basis to the money he is getting with the Yankees. What he couldn't get was the long-range security and "guaranteed" contract Kapstein wanted.

The parting was hardly amicable. And it may be a long while before the wisdom of the Yankees in taking the long-range gamble or the wisdom of the Reds in refusing it on an injury-prone pitcher will be resolved.

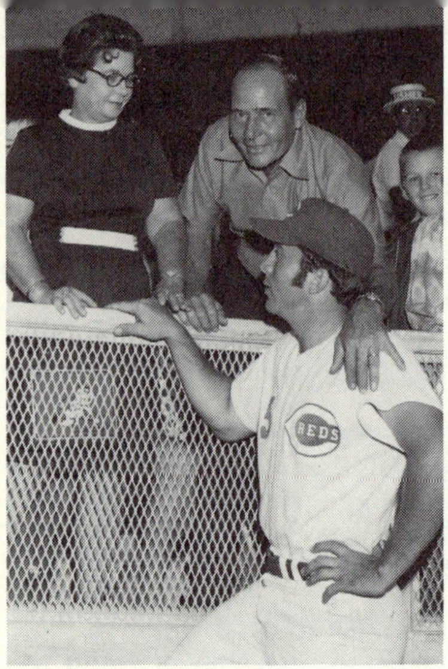

Katie and Ted Bench with their boy.

Celebrating Series triumph in Boston with chairman of the board, Louis Nippert

Johnny's human . . . one gets away

THE GLAMOUR BOY

The big Braniff jet was in its slow descent to land at Tampa International Airport on this muggy June evening in 1965. It may have been dropping gently over the International Inn, the resort motel that has been spring training headquarters for the Cincinnati Reds since 1962.

For Johnny Bench, an 18-year-old passenger from a place called Binger, Oklahoma, it was his first airplane ride. The big youngster with the broad shoulders and the massive hands had an uneasy feeling in the pit of his stomach — the sensation of uncertainty as one faces a lonely transition in his life.

In hundreds of future plane trips, the stewardesses would fall all over themselves to become acquainted with Johnny Bench, who got to know more of them than the president of any airline. But this night, Johnny Bench was an unknown, tense youngster.

He walked through the airport to the open air baggage claim area and then surprised a cab driver by asking to be taken to Al Lopez Field where the Tampa Tarpons were playing a home game in the Florida State League.

The Tampa Tarpons do not attract big crowds as a rule and the 500 or so persons in the stands had no inkling that they were to witness the debut of a young man who in a few short years would be one of the best known athletes in the nation. The game was in the third inning when he arrived.

Manager Merrill (Pinky) May sent his new player up to the plate as a pinch hitter. Bench, who had been the No.

2 player taken by the Reds in the then new free agent draft, grounded out then went behind the plate to finish the game as catcher.

"The next day they released the other catcher and we went on a road trip with me the only catcher," says Bench. "I didn't even know the name of the guy they dropped, but I felt sorry for him."

Within three years, the raw youngster from Oklahoma was in the majors and within five had become the youngest player ever to win the National League Most Valuable Player Award. A year after that, he had his own syndicated television show; had sung with the likes of Glen Campbell and Bobby Goldsboro, and was destined to head national charity drives, be invited to the White House, become the nation's most eligible bachelor and stamp himself the most popular baseball player of his time by leading the All-Star vote three consecutive years (1973-75).

Johnny Bench, the glamour boy who was to acquire an endless string of beautiful women, led a very sheltered life that first summer in baseball.

"I was barely making enough money to get by," he says, looking back, "and I was a pretty shy young man then."

That reflects his boyhood. There is no such address as Easy Street in Binger. It was and is a town of around 500 people. "If you see the 'Resume Speed' sign, you've missed it," quips Bench now.

Ted Bench's mother, Johnny's grandmother, was a Choctaw Indian. The Indian heritage is becoming more apparent in Johnny's features as he grows older. He has never dealt with it in any way other than of pride, although his fame makes it less of a point of interest now.

The Bench family was unquestionably poor. When

young Johnny displayed the early signs of being a fine athlete, it was a friendly benefactor who provided his first ball glove.

"We used to take turns knocking tin cans over the back fence with sticks" says Bill Bench, five years older than his famous kid brother. "Pet milk cans were our favorite, but sometimes they'd get pretty well cut up and we'd cut our hands playing."

Binger High School was too small to have a football team or Johnny Bench's fame might possibly have been as a star in another sport with the University of Oklahoma Sooners. But Johnny never learned the words to "Boomer Sooner" as a teenager and he played basketball and baseball at Binger.

In his senior year in high school, there was a bus crash bringing the baseball team home from a game and Johnny cracked several ribs, winding up in the hospital — a forerunner of frequent visits. As a matter of fact, he had a close brush with death in an automobile accident that first year as a pro, suffering head injuries.

"The first time I ever saw him, he was as bald as a billiard ball with a big scar across the top of his head," said pitcher Gary Nolan, recalling their first meeting in the Florida Instructional League. "He sure wasn't very pretty then."

"We had to work very hard for everything we had as kids," Bench said in an interview in his rookie season in the majors. "We did anything we could to earn a buck, mowing lawns, harvesting peanuts, loading trucks, anything. In fact, we'd start to school in August and have a two-week layoff in September so everyone could hire out to harvest the crops."

In later years, Ted Bench told an interviewer he used to give Johnny $5 for hitting a home run. His memory might be a bit shaky, because Ted didn't have surplus $5 bills in those days, but who wants to take issue with a proud father?

After establishing himself in Cincinnati and before he was making six-figure salaries, Bench purchased a motel on the north edge of the city and installed his mother and father as resident managers. A brother and sister also moved to Cincinnati. More recently, Ted and Katie have moved back to Oklahoma, but did not return to Binger.

"We were going to set up a hardware and general type of store in Binger," Johnny said in the spring of 1977, "but someone beat us to it. So we bought a home in Oklahoma City for the folks and we'll keep an eye open for business opportunities because my father doesn't like sitting around. As a matter of fact, he was doing some part-time truck driving over the winter just for something to do."

The statistics of Bench's beginning in Tampa would offer little clue to the greatness ahead. He hit only .248 and two home runs, but the raw material and strength were all too evident.

A year later he was promoted to the Newport News, Virginia level in the Cincinnati system and Manager May moved up the ladder with him. May, a onetime National League infielder, had a strong 15-year-old son who used to help Bench and the other catcher on the roster warm up the pitchers. He's Milt May, now with the Detroit Tigers and an established major league catcher on his own.

Bench was beginning to find himself in the Tidewater area of the Commonwealth. Among other things, he made a discovery that most young men find fascinating — girls.

"I was on a hitting rampage in August and I lost my virginity that same week," he told a reporter years later. "I was called up to the Buffalo farm club and fractured my thumb in the second inning of the first game I caught. I thought the Good Lord must be punishing me."

The Diety apparently did not take Johnny's subsequent amorous adventures all that seriously.

It has to stagger the imagination that an athlete of Bench's potential actually was passed over by all 20 major league teams in that very first free agent draft. The Reds exercised their first pick for Bernie Carbo, who obviously had major league potential and who eventually made it to the majors with the Reds under Sparky Anderson.

But Johnny Bench was the No. 2 draft pick of the Reds and the 35th athlete chosen in that draft!

To what do the Reds owe such a marvelous stroke of good fortune? Partly luck, of course. But in his senior year in high school, Bench was pitching and most scouting reports concentrated on his prospects as a moundsman. Johnny himself doesn't understand that. "I did hit .625 my senior year," he says.

The remoteness of Binger is another factor, but the Reds did have their projection on Bench as a catcher.

Broken thumb and all, Bench faced six months of military duty after the season ended in Buffalo and he wasn't able to be with the Reds in spring training in '67. He had only 11 days of pre-season conditioning before starting the season with Buffalo, then the top rung in the Cincinnati farm system.

One man in the Cincinnati organization watching Bench closely was Dave Bristol, who had become manager of the Reds in mid-season of 1966. "I only hope I'm here long enough to have him on my club," he said. He had made up his mind that Bench would be his catcher in 1968 and had suggested to the front office that John Edwards, the incumbent No. 1 catcher and an acknowledged pro behind the plate, could be traded.

"I didn't expect Johnny Bench to tear up the National League as a rookie, and he didn't," Bristol said. "But there was no doubt in my mind that he could be a very exceptional catcher and I knew him well enough to feel that there was nothing to be gained by giving him more experience in the minors."

Bench was called up to the Reds in September of 1967 and did all the catching, but his .163 batting average hardly seemed to justify Bristol's confidence. Nonetheless, the Reds did trade Edwards to St. Louis in February of 1968. One of the players obtained was Pat Corrales, a proven journeyman receiver who had problems hitting the curve ball. But he was an "insurance policy" for Bench.

Bristol used Bench in 154 games his rookie year, a staggering work load for any catcher. Johnny's hitting was a respectable .275 with 40 doubles. His output of 15 home runs was below his own expectations, but certainly satisfactory and adequate to help earn him Rookie of the Year honors.

Bench's fourth home run was a whistling line drive in Chicago's Wrigley Field, a cozy haven for homers. But this was still in an upward trajectory as it landed in the bleachers in left field. "I would like to have gotten that one up in the air to see where it went," Johnny said later.

Even as a 20-year-old rookie, Bench had his own definite theories about hitting. Explaining why pitchers with a lot of motion bothered him some, he said, "There's a world of difference in catching a pitch and swinging at it. The ball comes straight to you from the pitcher. You swing at the pitch from an angle."

In the spring of '69, the Reds had an exhibition game with the Washington Senators, managed by Ted Williams, the acknowledged modern master of the art of hitting. Bench asked Williams for an autograph. Williams took the ball and pen and wrote, "To Johnny, a Hall of Famer for sure" and handed it back to the surprised Bench.

The '69 season saw Bristol feel forced to discipline his star receiver in a "this hurts me more than you" figurative spanking. Feeling JB was acting a bit cocky, Bristol made him a spectator for three games in Philadelphia without making a fuss about it. The episode received scant publicity but Bench knew what the reprimand was all about.

The fine line between confidence and cockiness seemed apparent to Bench all along. When he became the club's No. 1 catcher, he immediately made it apparent by his actions that he wanted control in calling the pitches.

As the 1970 season rolled around, the best barometer of Bench's ability was the fact his third-year contract called for a salary of $42,500. In today's free agent market that seems trivial, but as recently as 1970 that was front-line player recognition. Bench had already established that he had a throwing arm as strong and accurate as any catcher who ever played the game. But while he was building respect as a power hitter, he had not yet proven his consistency against all types of pitching.

The Reds' front office was busily occupied with the details of getting into Riverfront Stadium. The original time table called for the club to start the season in the new downtown location which would nearly double the capacity of cozy Crosley Field, home of the team since 1912. The stadium was rushed to completion in time to assure playing the All-Star game, the one that attracted President Richard M. Nixon as a spectator and was won by Pete Rose's headlong crash into catcher Ray Fosse.

The last game ever played in Crosley was the night of June 24, with the Reds holding an astonishing nine-game lead in the Western Division. It was a nostalgia-filled night as an almost capacity crowd of 28,027 saw the game against the San Francisco Giants. Juan Marichal, the high-kicking pitching ace of the Giants, held a 4-3 lead going into the home half of the eighth inning. Bench, the first man he faced, was the owner of 24 home runs, only two under his previous season's total.

"I'm not usually a guess hitter," he said later. "But on this occasion I felt I was entitled to one, and I guessed fast ball on the first pitch. He had gotten me out on a high and away fast ball the time before."

Marichal did come up with a fast ball up and away but instead of popping it up, Bench got far around to send it into orbit — and it soared over the scoreboard in left center to tie the game. When Lee May, the "Big Bopper from Birmingham," followed with a towering hoist over the wall in center, the Reds had a 5-4 lead and provided a finish that couldn't have been more appreciated had it followed a dramatic script.

Bench assumed heroic proportions in every aspect of the game during that season, which found the Reds waltzing to the division title.

"Johnny Bench is as valuable to this club as any man in baseball could ever be, and that would include Babe Ruth," said Wayne Granger, the lanky relief pitcher, in a tribute that covered all bases.

"He's all over me when I'm not working as hard as he thinks I ought to be," said lanky Jim Merritt, the southpaw who won 20 games before tearing up his pitching elbow in September, an unfortunate event from which he never recovered.

Merritt's nickname in the Reds' clubhouse was O.C., which stood for Occasional Cocktail. The night he won his 20th game in Philadelphia's Connie Mack Stadium, there was champagne for everyone afterwards, purchased by the pitcher himself. Manager Anderson, who bars alcohol on the chartered flights, didn't object on this occasion and indulged in a swig of the bubbly himself.

The bubble burst for the Reds in the World Series against Baltimore. The Orioles, still humiliated and angry after their experience against the Miracle Mets the previous October, ripped the pitching-poor Reds in five games.

For Bench, however, it was a super season. At 22 he became the youngest player ever to win the MVP. His 45 home runs were the most ever hit by a catcher and the second highest ever by any Cincinnati player. His 149 RBI total was also a Cincinnati club record

Teammates Pete Rose, May, Jimmy Stewart, the ill-fated Jim McGlothlin (who died of cancer in 1975) and Jim Maloney turned out for the press conference announcing the MVP award. Naturally, Bench's family was there and Johnny announced he was joining Bob Hope on a tour of U.S. military bases in Vietnam and talked about his salary jumping to $100,000.

Johnny's commercials and TV appearances were beginning to amount to big money. And his reputation as a connoisseur of attractive girls was becoming established. His name popped up in gossip columns all over the country.

"I come in contact with a lot of stewardesses around the league," he responded to a question in 1970. "I meet a lot of girls. Other guys want to fix you up. It's not hard. There are girls everywhere and all you have to do is look around."

Bench's enjoyable whirlwind winter helped set him up for a monumental letdown. The defending National League champion Reds fell flat on their faces in 1971 for a variety of reasons, not the least of which was a drastic drop in productivity from Bench's bat. He lost 55 points off his batting average, hit 18 fewer home runs and lost 77 from his RBI total of the year before. This time there was no MVP party, but Bench's off-field popularity hardly seemed to suffer.

Manager Anderson responded with a "hard line" training camp in Tampa which most of the Reds accepted as fit punishment for their sins — Bench among them. As the Reds moved into the fourth hour of a strenous workout one day, Johnny yelled to his manager, "Hey, Sparky, there's a guy over here who wants to know where you want the lights installed."

One of the points Bench conceded in looking back on his 1971 off-season in productivity was that he had overlooked the fact that baseball was the cornerstone of his popularity. Translated into dollars and cents, he knew it was his home runs and success on the field that stimulated the other opportunities.

Johnny's 1972 comeback didn't quite equal his spectacular 1970 production, but it was close enough to earn him a second MVP. And he didn't need to apologize for an output of 40 home runs and 125 RBI. But the drama of that year for the husky young man out of Binger was still ahead at the end of the regular season.

Unlike their 1970 sweep over the Pirates in the championship playoff, it was uphill for the Reds all the way. The Pirates captured the first and third games of the five-game set. Going into the ninth inning of the final game in Riverfront they had a 3-2 lead. Dave Giusti, the righthanded ace of their bullpen, seemed a good bet to stave off the Reds. Giusti throws a "palm ball," a pitch that has downspin on it and breaks low and away from righthanded hitters. It is not a pitch to pull.

Bench was the first batter against Giusti and the script started to unfold in favor of the pitcher. But on a 1-2 count, Bench made contact with a breaking pitch and sent it into orbit to right field. Roberto Clemente merely watched as the ball reached the folded-up area of "football seats" behind the barrier, tying the game. The stadium exploded in a maelstrom of emotion and the home run was labelled "the most important ever hit in Cincinnati" by more than one writer covering the game.

"I had only one thing in mind," Bench said. "I wanted to tie the game and I was swinging from the heels. I knew he'd have to make a mistake for me to be able to pull the ball, but I was still cutting hard at it where he was pitching me. In his own mind, he didn't make a mistake. I was just fortunate enough to hit a real good pitch a long way."

The Reds won the pennant a few minutes later when pinch-runner George Foster tallied on Bob Moose's wild pitch, but then came the agonizing World Series loss to the A's.

The front office had scheduled a "Victory Party" that naturally turned into a drown-your-sorrows bash. Bench, not normally a heavy drinker, seemed morose and had more than his share of the bubbly, but he later went through an impromptu comedy skit with Dayton Journal Herald sports writer Bucky Albers.

Bench had more to be morose about than losing to the A's, but he had kept his secret well. In September, the Reds send their players through a routine physical examination. On Bench's chest X-rays, doctors detected a suspicious spot on his lung. Their first inclination was that it was an imperfect X-ray and they called him back. Again the spot appeared. Bench was called in a third time and the news broken that he would have to undergo exploratory surgery.

At Bench's request, the problem was kept confidential. Only his family and the upper echelon of the Reds' management knew the burden he was carrying through the playoffs.

The surgery on Dec. 9, 1972, confirmed that the lesion on the lung was benign. But even with that cheerful news, the big catcher had to carry another mental burden through the winter that he did not disclose to anyone other than that same inner circle.

"Dr. Luis Gonzalez, who performed the surgery, had told me that he would have to cut through some muscles in the area under my arms. He said that the surgical procedure should be adequate not to damage the use of those muscles. But he also said that as recently as 10 years ago, the surgery I had to have would have ended my career as a baseball player and prevented me from even swinging a golf club," Bench explained a few months later.

For the rest of his life, Bench will have a reddish surgical incision scar circling around his right side under the chest halfway across his back.

In the hospital he received 10,000 pieces of mail. His close friend, singer Bobby Goldsboro, called every day and sent a basket of fruit big enough to feed a couple of Little League teams. Bob Hope called him. Glen Campbell, Buck Owens and Chuck Conners sent messages. He had to stage a press conference in a hospital staff room and appeared in a bright red robe with a floppy, blue denim hat to the delight of TV cameramen and other photographers.

By 1972, Bench knew a great deal more about the workings of a television camera than just following the blinking red light. He had already made sufficient progress as an entertainer to be the host of a taped half hour show marketed nationally by an independent producer. His voice and delivery made him almost a professional singer in the country-western field and he appeared with Glen Campbell at the Las Vegas Hilton in November, just ahead of his operation.

Jim Marooney, an NBC producer, stood in awe as Bench walked away after finishing an interview with Tony Kubek the following spring.

"You don't find many athletes who can't give a decent interview any more," he said. "But Bench is very unusual in that he never misuses a word or abuses the rules of grammar. He has a total awareness of the tools of our trade."

Prior to his surgery, Bench had been addicted to golf and had been invited into the major celebrity tournaments, the Bob Hope and Bing Crosby affairs in California in January.

Bench took up tennis in the spring of '73, partially because of the operation and partially because there were always partners around the International Inn during spring training. All through that spring, he insisted he wasn't concerned about regaining maximum playing efficiency. But there were some inner doubts. He talked almost with a sense of destiny.

"I've been to the top, was up there real early, then I went down. People were booing me a great deal in 1971 and I reacted poorly to it at times. Then I came back. Facing surgery with concern over my future has had an effect on me. I think I'm a little more calloused about life in general now," he told an interviewer.

Permitted to pretty much set his own pace in Tampa, Bench was off to a relatively slow start in the season. He fretted as his average hovered around an anemic .200. But on the night of May 9 in Philadelphia Bench ripped three home runs off Steve Carlton, the classy southpaw of the Phillies. The Reds flew to Houston after that game to continue a road trip and the next morning a relaxed Bench opened a telegram. It was from Dr. Gonzales and it read:

"Johnny Bench's batting average up 30 points, Dr. Luis Gonzales surgery stock up 20 points. Congratulations on Philly performance."

On the season Bench's 104 RBI and 25 home runs was a fine year, except by the lofty MVP standards he had set in two of the preceding three seasons.

Johnny Bench, the human being, befriended two Dayton youngsters afflicted with leukemia, whom he met at an appearance on behalf of a department store. One of them was Philip Buckingham, who died in February of 1975 at the age of five.

In the short time that Johnny knew Philip, roughly a 14-month span, he made a number of trips to Dayton. The ensuing publicity helped Philip's family obtain support, both financial and spiritual, in their struggle to keep up the spirits of their doomed youngster and keep their family life afloat. When Bench became chairman of the Athletes Against Cancer, Carl and Melissa Buckingham had their first airplane ride to New York for the press conference. It was five months after their boy had died.

"Johnny and Philip had something super special," Melissa said at the gathering. "You would have had to see them to know what I mean. Sometimes it was like Johnny was the little boy and Philip the man."

When Bench talked about Philip having had his plastic Johnny Bench doll with him at the end, even most of the media had problems maintaining the detached, dry-eyed approach they are expected to display at all times.

If Bench's social awareness was on the rise, he also was coming close to the political field in expressing some opinions that reflected a definite conservative tone. His response to some Howard Cosell remarks about the Reds failure to catch the Dodgers in the 1974 pennant race set the tone.

"I'm tired of a negative press in this country," he said, "At a time when kids need heroes, the press is tearing everything down. It started with Cosell. Now everybody seems to be trying to earn a niche the same way."

President Gerald Ford had invited Johnny to the White House on at least two occasions and the chief executive took him on a personal tour on one of these. Bench defended President Nixon's attempt to sell his presidential papers, citing his own problems in the high tax bracket he found himself.

For all his brilliance as an athlete and his acclaim in other areas, the big catcher continued to live through a lot of stress. The year of his celebrated marriage to model Vickie Chesser — 1975 — was no exception. The storybook wedding in Cincinnati on February 21 was a "sellout" with a crowd of 1,000 persons milling around outside the church.

Vickie, an attractive model, had been seen in a number of television commercials, including a familiar one for a toothpaste that began with the jingle, "How's Your Love Life?"

Their courtship was a whirlwind affair, starting with a telephone call to Vickie in New York after Johhny had been given her number by a mutual friend. He impulsively asked her if she cared to fly to Las Vegas with him to spend New Year's Eve.

"I told him if he was crazy enough to ask me, I was crazy enough to go," she explained.

In his lifestyle and within the framework of contemporary attitudes, it wasn't a big deal in his eyes, but he conceded later he was surprised that Vickie had accepted.

She flew from New York to Cincinnati to join him and they continued on to Las Vegas. Their engagement was announced less than three weeks later.

Not many people remembered that Bench had been engaged in the winter of 1969-70 to an airline stewardess who lived in Cincinnati. That affair broke up before a definite wedding date was set when Johnny advised the girl he simply wasn't ready to settle down to married life.

"I was convinced that at age 27, though, I was ready," he said after the breakup with Vickie.

The wedding turned into the social event of the year in Cincinnati. At the rehearsal the minister expressed his concern over the possibility of media cameramen disturbing the dignity of the event. The invitations, starting with President Ford, were issued in a legitimate manner, despite some media inferences to the contrary, because Bench did indeed know the President, as well as the show business celebrities who were invited.

Attendance in the Christ Episcopal Church was by invitation only and a platoon of Cincinnati policemen kept a curious throng moving in front of the church itself. Security was equally tight at the reception in the main ballroom of the Netherland Hilton hotel. The tab, picked up by Bench himself, was more than $8,000.

Bench resented criticism of the fact invitees like Bob Hope, Glen Campbell, Bobby Goldsboro and others did not attend.

"All of them responded and we knew they weren't coming," he said. "The media didn't check back. They merely published the original guest list which we gave out when asked."

On April 22nd, in only the 16th game of the season, bridegroom Bench was involved in a fierce home plate collision with Gary Matthews of the Giants. He suffered a shoulder injury that ultimately posed a second serious threat to his career, although not in the same category with the cancer scare of 1972. A baseball career and life itself do not balance each other on any scale.

The pain persisted in his left shoulder and affected his batting swing, although there was no indication at all that his defensive play suffered. But a pessimism began creeping into Johnny's conversation that never had been expressed before. Even the ultimate satisfaction of being a key part of the ultimate goal of the Machine — winning the World Series — did not alleviate his concern about his future. Amazingly, he drove in more than 100 runs the fourth straight year and hit 25 home runs, in spite of the ache in his shoulder restricting his normal swing.

"If I had known in April what I was going to be in for over the season, I would have dropped out until the problem was resolved," he said after the successful surgery in which a piece of torn cartilage was removed from the damaged shoulder. Dr. Don O'Donogue had said the shoulder area looked as if someone had hit it with a hammer.

Prior to the surgery, he had shared speaking honors with Betty Ford, wife of the President, at a cancer luncheon. Johnny's frenetic schedule prompted a writer to claim without reservation, "No baseball player has come close to matching Bench's off-field popularity since Babe Ruth's heyday."

Amazingly there was no decline in that popularity during the breakup of his marriage to Vickie, which formally lasted a few days beyond a year but which in effect lasted about nine months. There has been a ton of "sob sister" reporting about the end of the marriage and there will be more. Vickie said some unpleasant things and called her estranged husband "a true tragedy as a person.'' Bench is willing to asume his share of the responsibility for the failure.

''I was the one who wanted out and I have to say I was at fault to a great degree,'' he admitted.

There were hints that all was not well in the marriage as early as mid-season. In an interview at that time, Vickie lamented that life couldn't be normal — as she saw it — in her relationship with her celebrity husband.

''We've never had one day since we've been married that we could be alone with nothing else to do,'' she said. ''We never had a real honeymoon.''

At first, during spring training, she had seemed to enjoy being interviewed. But the novelty of that wore off, as it did with the many travels and constant limelight. Vickie also discovered that Johnny has a way of keeping things very much to himself, as evidenced earlier when he concealed his impending surgery during the stress of the 1972 World Series.

Perhaps the basic cause of the split, apart from two persons being unable to adjust to each other's special needs, may lie in Bench's addiction to being where the action is and, more important, to being part of it. When he went to New Orleans to be king of the Mardi Gras in 1976 with his marriage in its final stages, he obviously wasn't making an all-out effort to change his ways in a bid to patch things up.

It's easy to think of the little boy from Binger who had nothing but dreams as a kid not yet having his fill of fantasies come true. On the other hand, most of the year the wind whistles and the dust blows around Johnny Bench Field, the Little League Park in his hometown, and where is reality?

The ambiguity of Bench's dreams also may lie in his continually changing his projected retirement date. When he was only 25, he spoke of achieving financial stability so he could quit baseball if he wanted to at 28. But when he was approaching his 28th birthday, the projected retirement had been pushed back a couple of years.

Now that his 30th birthday is approaching (December 7, 1977), retirement wouldn't appear to be in the immediate future.

Bench's special status in the minds of baseball fans was never more evident than in 1976 as the Reds were enroute to a second straight World Series. By any standards, Bench had an off-year with a mere 16 home runs and a .233 batting average. Yet he finished strong enough in the sweep over the Yankees to be named MVP for the Series. And in his opinion, his shoulder never has come back to full strength.

But even if he never played another game, the part-Indian boy from Binger is Hall of Fame material — with few worlds left to conquer before his 30th birthday!

"You asked about motivation," he said to a reporter in spring training of 1976. "All my life I've been faced with living up to the expectations of other people. I don't think things are going to change much."

Here comes Fred's "screwgie"

THE ACROBAT

Fred Norman's early log in baseball reads like one of those souvenirs of the American past — an authentic railroad passenger timetable. But then no line, not even the legendary Santa Fe, could have hauled Norman around from Shreveport to Binghampton to Lewiston to Kansas City to Fort Worth to Wenatchee to Chicago — and beyond.

Fred Norman changed uniforms 19 times between signing as a hard-throwing bonus prospect with the Kansas City A's in 1961 and his return to the majors to stay with the San Diego Padres in 1971.

The span of three-plus seasons Norman put in with the Cincinnati Reds going into the 1977 season was by far and away the longest stretch likeable Freddie has put in anywhere in his pro career.

"I found a home and I like it," he says.

There is no record of Norman doing a back-flip upon hearing of his being traded to the Reds on June 12, 1973, although he was known for that feat among his buddies on the Padres. His most celebrated public back-flip was off the mound in Dodger Stadium on a September night in 1971 when he beat the Los Angeles club to eliminate them from the Western Division race.

"It was sort of a dare," he laughed. "I had been doing the back-flips while we pitchers were running our laps. Somebody asked me why I didn't do it when I won a game. I said I'd do it when I beat the Dodgers. You know the two teams are pretty big rivals and on this occasion San Diego had lost to them 10 straight times. I never had any great love for Walter Alston (the Dodger manager) because I felt he didn't give me a fair chance when they had me."

As far as Norman knows, his back-flip wasn't recorded in the Los Angeles press, which was more concerned with the burial rites of the team.

The trade that brought Norman to Cincinnati wasn't exactly a deal. It was a cash auction. Buzzy Bavasi, the Padres general manager, was strapped for working capital in the summer of 1973. C. Arnholt Smith, the team's majority stockholder, was having problems with the federal government as he tried to save his banking empire from collapse.

Ironically, the Reds tried to get Bavasi to put a price tag on Clay Kirby, a younger pitcher who had the knack of beating Cincinnati with great regularity. But Bavasi didn't want to yield Kirby and he also insisted that the Norman deal be disguised. The irony is that the Reds did trade for Kirby a few months later, only to find him a tremendous disappointment whereas Norman is still going strong.

Because Bavasi disguised the sale so as not to offend the faithful Padre fans, however few they may have been, the Reds sent outfielder Gene Locklear and a minor league pitcher along with the six-figure check.

Locklear is a full-blooded Lumbee Indian out of western North Carolina who wields a big league bat but has been a defensive liability. He is also an artist of considerable talent.

Norman was not taken by surprise at the deal.

"Buzzy had told me," he explained, "that they were in the process of selling my contract to meet some obligations. In fact, he let me know which teams were interested and asked me if I had any preference. I told him I'd prefer to stay in the National League and I knew the Reds were having problems, and he accommodated me by working it out that way.

"Later, after Ray Kroc (founder of the McDonald's Big Mac empire) came along and purchased the team, Bavasi said one night in a television interview that if Kroc had acquired the club sooner he wouldn't have had to give up a pitcher of Norman's caliber. I thought that was nice of him."

Realist Norman felt he was ready for the Reds at that time on two counts. The first was a rebirth of his screwball during a brief stay in the minors in 1971 and second his gradual emergence with the Padres.

Freddie's family was on the move through the 1930's in the wake of what historians call the "Great Depression." They were living in San Antonio, Texas, when Fred was born in 1942. But by the time Fred was eight, his father was running a small restaurant in Arcadia, Florida, a crossroads town inland and a bit north of Fort Myers.

"He had always dreamed that one of his boys would be a big leaguer, and I was the fifth and youngest. He had pitched some in his youth but he was 47 when I was born. He decided we would move to the Miami area so I could compete in a better level of high school competition and scouts could get a better look at me."

It was one of the few times in his life that the senior Norman made a move that paid a big dividend. His stocky son didn't grow very tall but he was a strong kid who could fire a baseball with real gusto. He had 12 offers to sign a pro contract when he graduated from high school.

If the Norman family had gone through a version of Woody Guthrie's "Bound for Glory" in its travels in the 1930's, Fred's meanderings through the 1960's were filled with a different sort of frustration, so much so that Fred concedes the temptation to quit more than once.

"It isn't really easy to say that you came close to quitting, but it happened," the little southpaw says. "I had a lot of frustrating things happen to me. What it seems to come down to is how many times can you get up after being knocked down? After so many times, you tend to ask what's the use. But I kept bouncing up and I'm very thankful that I did. I always liked baseball but I've learned now to really love it."

Norman's frustrations included arm problems, personality conflicts and other complications.

"But there's always the little incentive of asking yourself what you've contributed to the game or yourself. In my case, I had to prove something to a whole lot of people who said I was too little."

Particularly unrewarding were Norman's four seasons with the Cubs. Among other things the Chicago braintrust told Fred was to forget about the screwball he had developed. Then the Dodgers got him and he divided the 1970 season between the Los Angeles club and the St. Louis Cardinals, who ultimately dropped him to their Tulsa farm team where he got an unexpected but very meaningful break.

The manager at Tulsa turned out to be an experienced lefthanded pitcher named Warren Spahn, the 300-game winning Hall of Famer. He took one look at Norman's screwball and immediately suggested he restore it to his repertoire.

"When a man with Spahn's record tells you your best pitch is good enough to win in the majors, it really does something for your confidence," Fred recalled. "The best part of it was his vote of confidence came at a time I knew I had to be losing a little bit off my fast ball.

"The Cubs had me tagged as a guy who couldn't make pitches where he had to make them. In other words,

I was wild, and when you get a reputation like that, it's hard to get rid of it. I think that hurt me with the Dodgers. The man (Walter Alston) didn't have confidence in me. He was afraid I'd walk guys in the clutch.

"That's the beauty of going to a team like San Diego. They needed bodies and they'd take the trouble to find those things out for themselves."

So the Padres rescued him and his "screwgie," which pitching coach Roger Craig helped him refine.

"Roger is a terrific teacher," Fred says of the lanky one-time Dodger who beat the Reds in his major league debut way back in 1955 in Brooklyn's old Ebbets Field. "He taught me what pitches to use to set up the screwgie and how to work against certain hitters.

"San Diego was an absolutely perfect place for me to finally get my feet on the ground. They were only in their third season and the opportunity definitely was there. I was confident I could win in the majors and the only pressure on me was what I put on myself. Still, it could be a depressing place because we had very few big crowds. Most of the time it was 5,000 or thereabouts and you'd hear the same sounds night after night."

But Norman also grew to love San Diego as a community with its blend of beaches and mountains and consistently pleasant temperature. An outdoors type, he found plenty of things to do away from the park.

Realist Norman, however, was ready when the opportunity to go to Cincinnati came. At the time he was wearing a Fu Manchu beard and his long flowing hair flopped crazily from under his cap. Curiously, his 1-7 record showed the only victory a 2-1, six-hitter over the Reds on May 20.

The long hair and facial adornment were gone when Norman walked into the Reds' clubhouse in Riverfront to greet Sparky Anderson.

No pitcher, be he Lefty Grove or Sandy Koufax or Spahn, could have walked into the situation in Cincinnati and made a more spectacular debut than Norman. With Gary Nolan on the shelf, the entire Cincy pitching staff was jumbled into one monumental slump.

Fred faced the Pittsburgh Pirates in his debut and pitched a route-going shutout. Next came the San Francisco Giants and another shutout by the little southpaw. His third start was against his old friends, the Dodgers. Ron Cey, the third baseman who answers to the nickname of Penguin, hit a ninth inning home run, but all it did was deprive Norman of his shutout. But it added up to three complete games and 26 scoreless innings.

Norman will not argue that he was a trifle over his head. "But I was fired up, tingling with excitement, coming to a club that was going to be in the division race. It was a helluva way to break in."

He was 12-6 over the season after joining the Reds, which left him with an overall 13-13 record and his competitiveness made him more than welcome. Still, acceptance as a regular in the rotation was slow in coming.

Even with his contributions to the cause and five strong innings in the game he started against the Mets in the stormy championship series of 1973, Fred did not find himself singled out as a certain starter in spring training in '74.

"I shouldn't have been surprised, and I guess it's the same old thing. If you're running the show, you're looking for someone better than a little guy who has turned 30. But they always have come back to the little guy here and he delivers," Norman said.

Freddie had nine fewer starts and pitched 54 fewer innings in 1974 and wound up with a 13-12 record. In terms of what he had done in a Cincy uniform the year before, that represented a dropoff.

Again in 1975, Fred wasn't in the rotation as the season opened and was sitting on an inconsistent 3-3 record the night in June that Don Gullett suffered the fractured thumb on his pitching hand. Moved immediately into the rotation, the veteran lefty responded by winning nine of his remaining 10 decisions and finished with a 12-4 mark.

Skipper Anderson gave Freddie the biggest vote of confidence he's ever had by designating him to pitch the second game of the post-season series against the hard-hitting Pirates. In nominating Norman, his manager gave an insight into a criticism that has haunted Freddie all along.

"When he goes out and challenges the hitters, he's a fine pitcher," Sparky philosophized. "When he's out there experimenting, Freddie the Scientist worries me."

Norman took that message to heart and fired out "smoking" his best fast ball and screwgie at the Pirates until he was starting to lose a little velocity. Then Anderson called Rawly Eastwick, his youthful cool customer out of the bullpen, to finish what was a 6-1 triumph and Fred's first post-season victory.

The surge to the second championship in '76 found Fred again not getting into the rotation until problems set in. Again he delivered, including a good effort (12-7) and a no-decision start against the Yankees.

Going into 1977, Norman had a new incentive. An attractive wife named Cindy is to be on the sidelines and the longtime guitar-strumming bachelorhood era is ended.

"Cindy is considerably younger than I am, but I know she's right for me," Fred said of his bride, then only a year out of high school. "She just happened to be the one who came along who was right for me.

"I've had a funny reaction in spring training," he said in Tampa "I've had to tell myself not to throw too hard trying to impress her. I caught myself watching her when I was throwing batting practice the other day. I've noticed other pitchers' wives who get awfully nervous when their husbands are pitching and I've seen them get up and leave their seats when he's in trouble. Cindy and I talked it over and she said we'd better get used to each other in that respect because she's not going to stay home when I'm pitching."

Norman viewed the departure of Gullett as something that could help him over what he considers the final hurdle of his career.

"I feel that I've established myself as a major league pitcher up to a certain point," he says. "What I'd really like to do is establish that I can win big. I've got a few more years left in this game to enjoy.

"As I look back, I always seemed to be used about three months out of a six months' season. I know it's not too late for me to win big. I don't give up easily. I love being with the Reds and I'm not criticizing anyone, but I'd certainly love to start a game in the first series of the season, and go from there."

THE SMOOTH ONE

If there ever is a Spanish-speaking version of the Reds radio network, it would be a rank injustice if someone other than Senor Juan Vene of Caracas, Venezuela, turned up as the play-by-play announcer. In fact, some listeners of the Venezuelan's sports broadcasts may think he already is in the employ of the Cincinnati club.

Somewhere in the past the affable Senor Vene adopted the Reds as his very own. When he visited this country several times a year to do special baseball broadcasts he spent as much time as he could around his favorite team.

Juan's loyalty to the Reds contributed in no small way to David Ismael Concepcion, the ever-so-smooth shortstop, coming into the clutches of the Cincinnati organization in the winter of 1967-68. The 19-year-old Concepcion already had attracted much attention in his home town of Aragua as a pitcher. Then a baseball-infatuated gentleman named Jaime Caribe hired young David to play on a team sponsored by a bank he owned.

So it was that David moved to Caracas to work in the bank when he wasn't pitching or playing another position on the baseball team, which was involved in the Venezuelan National Series. Anyone who has seen Concepcion's strong arm — a "cannon" in manager Sparky Anderson's description — wing down a runner from deep in the hole can appreciate the fact that he wasn't all that bad a pitcher.

Senor Vene knew all the major league scouts who

were congregating in Caracas for the national games and had convinced David's "protector" — the banker — that their young star should go to the Reds. David himself, however, recalls that he was open-minded on the subject.

"Some of my friends talked to me and said the New York Mets were willing to give me a good offer," he said. "But they hide me and take me away from my house and the man from Mets could not find me."

The "they" was the banker. And so it happened that Senor Vene could direct Wilfredo Calvino, then the Carribean scouting supervisor for the Reds, to the place where young David was staying. It was no problem then to sign him to a minor league contract in the Cincinnati organization.

Baseball had always been in his mind as David grew up. Luis Aparicio, the longtime star shortstop of the Chicago White Sox, was a Venezuelan national hero. A graceful, quick artist with the glove, "Little Luie," as they knew him in Chicago, played some in Venezuela during the winter.

"I see him play and I say to myself, 'I can field like that too' and I like to play shortstop," David recalled.

David's father was a truck driver who owned his own dump truck and eked out a living hauling whatever he could for whoever needed his services. His mother was a tailor who made clothes and sold them to help with the family finances.

Once he started earning major league pay, David became the main support of his family. He began sending them a check for $150 out of his twice-monthly paycheck. Considering that he was making only $13,000 in his second season in the majors, that was no minor outlay for him.

There is a story that Concepcion's father wanted him to study to be a doctor, a story that amused his Cincinnati teammates who know David becomes unnerved at the sight of blood. In 1971 pitcher Jim McGlothlin was struck in the face by a hard smash off the bat of a San Diego Padre hitter and crumpled to the ground with blood gushing from the right side of his face. Tommy Helms, then the Cincinnati second baseman, tells the story.

"We all ran over and the trainer was trying to stop the blood with a towel and had already waved for a stretcher. It wasn't funny, but I almost had to laugh at Davey. He had turned white, he stood there moaning and I thought he was going to faint. He'd sure made some doctor!"

Concepcion is vague about the wishes of his father for him to enter the medical profession and allows he had little interest in extended schooling.

"My sister, she became a teacher, and my father wanted me to stay in school," David said. "I started to what you call high school, but after one or two classes each day, I leave. After three weeks, I quit and go to work. I do not regret that decision because I was able to play baseball and get the job in the bank in Caracas. I liked the job in the bank too, and then I come to the United States. Baseball has been good to me."

When David reported to Tampa to make his pro debut in 1968, he encountered the fate of many a youngster moving up a notch in the game. He was over-matched against the caliber of pitching he faced.

"For six weeks he was helpless with the bat in his hand," recalls George Scherger, the manager. "But he stayed with it and didn't let it affect his fielding. He was supposed to be a second baseman when I got him, but I had a kid at second I regarded as a good prospect and played David at shortstop."

The Reds were expanding and revamping their minor league training system at the time and Ted Kluszewski was brought back into baseball as a part-time batting tutor. Klu's first class in the Instructional League that October included David, who had a horrible time trying to pronounce Ted's name. To this day, Concepcion does not pronounce the "K."

"I liked his stroke even then," Kluszewski recalls. "But he was a real wiry kid then and I had no inkling he would fill out physically as he did. Even now he fools you with his arm strength."

All told, David has had only two seasons of minor league experience, dividing the '69 campaign between Asheville and Indianapolis.

The Reds were well fortified with shortstops at that time and the one with the highest initial promise was Frank Duffy, out of Stanford University. Then Darrel Chaney, who had been the No. 2 choice behind Gary Nolan in the 1966 free agent draft, had finished his minor league development. The veteran Woody Woodward also was available.

Still, new manager Anderson was greatly impressed with Concepcion as he put his first Reds team together.

Ed Liberatore, a veteran scout who had once worked for the Reds but by then was with the Dodgers, told a writer:

"The Reds are crazy if they think they have anyone who can play shortstop better than that Venezuelan kid."

Anderson did some fancy shortstop juggling in his first season with Concepcion making it into 101 games, although a few of those were as a late-inning defensive replacement. He hit a respectable .260 and seemed on his way.

Nothing at all seemed to go right for the Reds in 1971 and Concepcion had his bad break early — literally the fracture of the thumb on his throwing hand. Rich McKin ney of the Chicago White Sox slid into him in the spring's very first exhibition game putting him out of action for five weeks, into the second week of the regular season. In the general demise of the Reds, his average dropped off to a disappointing .205.

Many of Concepcion's teammates regarded him as a "hot dog" at this stage of his career, and there was even a hint out of the front office that he might be demoted to Indianapolis to provide some badly needed motivation. Sparky Anderson preferred, however, to more or less turn the youngster over to Tony Perez, the "big daddy" of the Latin athletes and a man of great sensitivity in motivating and riding herd on the others.

David married his longtime sweetheart, Delia Montenegro, before reporting to spring training in 1972, and this union eventually helped him mature. In '72 David's salary was up to $20,000, plus World Series money that fall. He and his bride used most of that Series loser's share to purchase a condominium in Maracay, where they still live. David also was beginning to make some modest business investments, relying mainly on the advice of his banker friend with whom he has retained a close relationship.

Concepcion launched his fourth season in the majors in a blaze of batting fireworks. He actually was leading the league after 10 games with a .414 figure and his defensive play was now recognized around the league. His manager, Sparky, delighted him by naming him to the 1973 All-Stars as one of the extra infielders.

Tragedy struck, however, the Sunday before the All-Star game. The Reds were playing Montreal in Riverfront and had a 6-0 lead in their half of the sixth inning. Concepcion lined a base hit, the last he would be getting as it turned out, and was on first base. Denis Menke chopped a slow roller to the left of the mound and was thrown out on a close play at first.

Concepcion never broke stride rounding second and was trying to take the extra base on the infield out. As he slid into the base, the spikes on his left shoe caught the bag and the impact caused a fracture of his ankle. Coach Alex Grammas waved wildly to the dugout for help as Concepcion beat his fists into the artificial surface in a blend of frustration and pain.

Later in the subdued clubhouse, Pete Grammas, a 9-year-old twin son of the coach, blurted out, "It's a good thing it wasn't Pete Rose."

Grammas, who left the Reds after the 1975 season to manage the Milwaukee Brewers, has contended all along that Concepcion is the finest defensive shortstop he's seen in his years in the game.

"He can make plays that nobody else I've seen could," insists Alex, who was in charge of the Cincinnati infielders.

Another indication of how far Concepcion had come at the time of the injury was the reaction of the man who would replace him in the lineup.

"In the spring, I thought I could beat him out for the job," said Darrel Chaney, "But he's the best shortstop in the National League now by far. I'll be out there doing a job but let's not kid anybody. I'm no Davey Concepcion."

Concepcion was bitterly disappointed a year later when Yogi Berra, the All-Star manager, did not name him to the team.

Perez continued to be a strong influence on his younger friend, but David still was capable of impulsive action and then not being able to cope with it. One day in St. Louis he purchased a "mod" suit splashed with large brown and white squares. When he wore it on the team bus it was the immediate target of numerous jibes, action typical of the Reds whose trademark has been the way they "needle" each other in the clubhouse and around the batting cage.

"You look like Bozo the Clown," snorted Pete Rose, who didn't suspect that David really was very upset. The sulking Venezuelan had a bad night at the park against the Cardinals and never again was seen wearing the suit.

Even today there is a great deal of the 'little boy" in Concepcion — who says he hopes to work with youngsters after his playing days are over.

The ankle injury in 1973 prevented Davey from playing winter baseball for the first time. He battled boredom by taking up a hobby that would have caused shock waves of consternation in the Reds' front office had anyone been aware of what he was up to.

What he was up to was dirt road racing, Venezuelan style with scant supervision. He purchased a Volkswagen and spent a considerable amount of money fixing it up. He installed a new engine in the six-year-old vehicle, removed the windshield, added roll bars, trimmed off parts of the fenders, installed shock absorbers, bought helmet and goggles and went racing.

"In my first race," he said, "they gave me three practice laps. I was nervous and ran over the wheels (rubber tires bordering the turns) three times. But they start me last in the 16 cars. I finished fourth, I passed eight, the others broke down."

Concepcion's investment in his racing equipment apparently gave him the same kind of competitive edge over his Venezuelan rivals that A. J. Foyt's superior equipment enjoys at the Indianapolis Speedway. He won three of the next four races.

Sparky Anderson gulped when David showed him a

color picture of himself with a mustache, proudly sitting on the hood of the VW. The words "Piloto David Concepcion" and a Pepsi Cola symbol were painted on the door.

When David played 160 games in the 1974 season all doubt had been erased as to his ability, not only afield but with the bat and as a base runner.

"You will be voted by the fans the starting All-Star shortstop next year," his manager assured him.

And indeed he was — both in 1975 and 1976. He also earned "Golden Glove" honors as the standout defensive shortstop for three consecutive years.

Odd things continue to happen to him. On June 6, 1975, a radio station in Caracas carried a news bulletin that he had been shot to death. The telephone wires buzzed between the northern and southern hemispheres as Concepcion assured family and friends that he was alive and well in Cincinnati. When things calmed down, Johnny Bench quipped:

"They heard how he got picked off base in Pittsburgh the other night, and they shoot people down there for that!"

In his native Venezuela, Concepcion is enjoying the stature of a national hero as Luis Aparicio once did.

His 1976 play with the Reds found him adding a new dimension to his skills, that of clutch hitter. On a team that had such feared RBI producers as Bench, Perez, Joe Morgan and George Foster, Concepcion led the club with 15 game-winning hits. More dramatically, he delivered five game-winning blows in one-run games.

His .281 regular season average was his most consistent. "David has just enough power to tempt him to go for the long ball on occasion," manager Anderson warns. "That's one thing we have to watch."

"I do not know if I will ever hit .300 but I would like to

and I think I can,'' Concepcion says.

Often overlooked is the fact that many an outstanding shortstop has never come close to hitting in David's range and still has been a valuable asset to his team. He will not be celebrating his 30th birthday until June of 1978 — so time is on his side to build his legend.

''Maybe somewhere there has been a man who played shortstop as well as he does, but I assure you there has never been a man who can cover the amount of ground he covers,'' insists Sparky.

The young man the banker and the broadcaster steered toward Cincinnati is in a class by himself.

Yes, Ed Kranepool, Davey got you

Classic DP ballet over Mets' Tim Foli

Rawly, Will happy world champions

THE ARTIST

It was the ninth inning as the Reds and New York Mets battled July 24, 1975. Smallish Fred Norman held a 2-1 advantage over Tom Seaver as the Mets came to bat in the home ninth.

With a runner in scoring position and two out, rangy slugger Dave Kingman was striding to the plate. A home run would win the game, a single would tie it. Manager Sparky Anderson waved to the bullpen for Rawly Eastwick, his still relatively unknown young relief pitcher.

The deceptively slender Eastwick, who acts as if he doesn't have a nerve in his body, cooly eyed the menacing Kingman. He zipped a fast ball across the outside corner of the plate just above the batter's knees for strike one. The kid came back with another fast ball, this one about three inches higher, which Kingman missed with a wild swing. Eastwick's third pitch was still another fast ball, this one letter high which Kingman also swung at and missed.

Rawlins Jackson Eastwick III had just given Kingman the classic "step ladder" treatment.

That demonstration was a tipoff as to Eastwick's approach to his job of chilling enemy threats in the late innings. His methodical manner and deadly results led to a writer calling him "Sparky's youthful iceman."

Anyone with the name of Rawlins Jackson Eastwick III sounds more like a student of the classics than a professional athlete — and Rawly is a little of both, being an artist by avocation and coming from an unusual family background.

There is no aristocracy in the family, according to Rawlins Jackson Eastwick II, who in 1977 was about to retire from his job with the New Jersey Bell Telephone Company and complete his work on a college degree. Ruth Eastwick, who along with Rawlins II raised five youngsters including two sets of boy twins, is finally taking formal lessons in painting.

RJE the first was named for a Rawlins who had been a general on the staff of U.S. Grant in the Civil War.

The elder twins in the Eastwick family, Bob and Dick, both were football players at Temple University and are high school coaches and teachers. Ralph, the twin of Rawly, has a college degree in political science, but in the spring of 1977 was associated with an insurance company.

As for Rawly, he is thinking of a long career in baseball but perhaps an even longer one in art when he reaches that point in life. If you suspect RJE the third marches to a different drummer than your average athlete, you are right. A bachelor, he has purchased an old house in Cincinnati and is restoring it to his own artistic tastes.

"It's an old Tudor house with pretty big rooms and high ceilings, just perfect for what I want to do," he said in spring training of 1977.

"I had all the carpeting torn out on the first floor and the floors sanded. I'm putting a fireplace in the master bedroom on the second floor. There's a big solarium in back that I'm turning into a studio for myself. I've got a guy redoing the kitchen with all new appliances. It's really kind of involved, but after it's all done, it's going to be my place, just the way I want it."

Rawly the artist does not think his interests are incompatible with a career as an athlete.

"My life in baseball is different than my life off the field and I intend to keep it that way. I like the excitement and challenge of the game, but when I go home I'm in a different world. That's why my house will mean so much to me. It's going to be all me."

Rawly's early interest in art was due to his interest in the landscape paintings his mother did.

"The oils and brushes were all around the house," he explained, "and one of my older brothers dabbled at it, but I got deeper into it eventually than he did.

"I took art classes in school, but they didn't mean much to me because I really didn't learn anything about artist's lives or techniques that I didn't already understand."

Eastwick says he doesn't remember being kidded as a high school athlete about his interest in art. And as for the direction he thinks his interest is going to take, he admits to a degree of confusion.

"It's getting stranger and stranger," he admits. "I like abstracts and I like to make up things and put them on canvas. But up to now, I've painted a lot of still life and landscapes. The truth is I've fallen behind the last couple of years because of baseball and the off-season demands on my time. I'm thinking about some formal lessons in portraiture.

"I have a mind that likes to explore all the avenues and I may want to express something social or political in my art. But mostly, I'm addicted to it because it is relaxing. I'll come home from a ball game and pick up some pastel crayons and just do a few simple designs to relax."

Despite the admitted thoroughness of the Cincinnati organization, Eastwick's conversion to relief pitching was not part of a grand design.

"I was 0-6 as a starter at Tampa in 1970 and all of a

sudden I found myself in the bullpen," he recalls. "I started to develop a slider under Dick Kennedy the next year," he added, referring to a long-forgotten manager in the Cincy system.

The full commitment of Eastwick as a relief specialist didn't come until the 1972 season when he appeared in 66 games without a start at Three Rivers and enjoyed his first productive campaign. He was 9-9 with a 2.34 ERA.

Then came two seasons at Indianapolis in which he was an occasional starter but was primarily the club's man in the bullpen.

"I thought I had the Reds made in the spring of 1975," Rawly recalls, "and it was a heckuva shock when Sparky told me I had to go back to Indianapolis. But he promised me that I would be back with the Reds shortly, that they simply didn't have enough work for me in the early weeks with open dates and Clay Carroll available.

"I wanted to believe him but I wasn't all that sure. I did real well in 13 games at Indianapolis and was called up. As a matter of fact, I got into Cincinnati about 3 o'clock in the afternoon and was in a ball game that night."

As the season progressed, Eastwick and young Will McEnaney were so effective out of the "pen" that they elbowed the veteran Carroll out of the spotlight. The performance against Kingman is typical of the way Rawly challenged hitters.

"I like the idea of coming into a difficult situation," he puts it. "I'm confident I can throw strikes. I never have a negative thought."

After the All-Star game in 1975, which found the Reds building an ever-widening gap against the Dodgers, Eastwick allowed only six runs in 47 innings.

Rawly's work in his first World Series shows a 2-0

record, but he remembers vividly a pair of home runs he gave up that added greatly to the excitement. For example, if he hadn't surrendered a two-run homer to Dwight Evans in the ninth inning of Game Three, the Reds would have won 5-3 and the controversial "interference" episode on Ed Armbrister's sacrifice bunt in the 10th inning never would have taken place.

Even more dramatic was the three-run homer Bernie Carbo ripped off him to tie the sixth game at 6-6 at a time the Reds could almost taste the victory champagne they were forced to wait another night to savor.

"The Carbo home run shows you what a fine line there can be between success and failure," Eastwick says. "I came so close to striking him out on the pitch before. He just barely got a tick of the ball and fouled it. He was badly fooled and way off stride. Then he came back and hit a real good pitch a lot farther than I ever dreamed he would considering the location of the pitch."

Eastwick was off to a relatively slow start in 1976, blaming it on the late opening of the spring training camps.

"I went into the season with the very strong desire to win Fireman of the Year honors," he said. "It was my reaction to people who asked me about the sophomore jinx and things like that. I didn't feel strong or consistent early but then things fell into place."

Falling into place is an understatement. He posted an 11-5 record, notched 26 saves (four more than he had the year before) and appeared in 71 games. He gave up only three home runs all season, which was a rather effective response to any suspicion that he might develop a weakness out of what happened to him in the World Series.

Oddly, he had another lapse in the championship series against the Phillies, giving up the runs that put the Reds behind 6-4 going into the ninth inning of the third game. But the Reds came storming back to win it and Rawly was the winning pitcher, despite being charged overall with five runs in three innings.

It seems incredible that the Reds swept the Yankees without calling on their ace fireman.

"It just worked out that way," manager Anderson put it. "I always knew he was back there, but in Yankee Stadium, the situation in the late innings called for a lefthander."

Eastwick's theory on pitching isn't at all complicated, nor does he intend to make it so. "I hear pitchers talk about concentrating on the weakness of the batters. That's fine, but to me, I'm going to challenge him to some degree with my best, which is a good, sinking fast ball.

"I'm much more effective going out and blowing people down than I am trying to finesse them. It's much easier for me to walk out there thinking I'm better than they are than trying to figure out how I am going to fool this guy."

Eastwick, at 6-foot-3, 185 pounds and boyish face, does not look the part of a power pitcher. But baseball strength is deceptive. The strength of some of the best hitters the game has known, men such as Stan Musial, Hank Aaron and Ted Williams, came from strong wrists and timing rather than raw, physical power.

Rawly's wrists are large and flexible, and his right forearm shows very little tapering in size down to the wrist.

The better the hitter, or perhaps the better the hitter's reputation the more Rawly's competitive juices flow. He especially enjoys a confrontation with Mike Schmidt, the powerful Philadelphia third baseman who led the National League in home runs in 1974, '75 and '76.

"I come from the Philadelphia area and there is an

aura about Schmidt to the fans there,'' he states his feelings. ''It's a real challenge for me to pitch to him and I've enjoyed some success. I give him inside heat and sliders away. He's a very strong guy and he can hit the fast ball out to the opposite field, especially in Philadelphia.''

Schmidt returns the compliment. ''He's one helluva tough pitcher for me,'' says the slugger, who was born and reared in what is Reds Country — Dayton, Ohio.

Eastwick also savors competition against Steve Garvey and Ron Cey, the power boys of the archrival Los Angeles Dodgers.

''I don't ever intentionally give Garvey anything over the inside half of the plate,'' Rawly says. ''He's very strong and very tough to fool. Every so often I'll challenge Cey with inside heat, but not Garvey.''

Sparky's ''iceman'' says he goes through no psyching rituals in the bullpen during a game.

''I just stay as calm as I can,'' he explains. ''As I've said, I am two persons. In uniform I like to work up excitement when I'm called to warm up, not before. Off the field, I'm getting to be a more quiet, introspective person.''

Eastwick has come to the conclusion that he wastes an awful lot of time in his travels, an analysis that seems to occur to only a mere handful of athletes. Many of them sit around the hotel all day doing nothing until it is time to go to the park and some go through careers visiting cities such as San Francisco, New York, Chicago and Atlanta without ever getting the real feel of those communities.

''I like to read, and I'm also going to start taking a few basic paints and small canvasses on the road, and keep a low profile,'' he said.

''To me, it's not worth going out and getting mixed up with a lot of people, and my social life in Cincinnati is going to be a little different with my home and my new girl friend."

Baseball players, especially those bachelors as attractive as Eastwick, have no difficulty meeting and dating girls. Rawly, who has savored his share in his brief major league career, has different thoughts going into the 1977 season.

"I've got a girl friend now that I enjoy being with and I don't see any reason to be spending a lot of time with other people, partying and things like that.

"I'm an admirer of Paul McCartney, and I've read his thinking about the fact he was caught up in this way of life, running around with all these girls and having a good time and then he met one woman he wanted to spend time with and he changed his life style. That's what is happening to me, and I find much more satisfaction in doing it that way."

The horizon, in both life and baseball, seems bright for RJE the third. He seems to have a good grip on himself and well mapped out directions.

THE BIG DOG

It is roughly 4,000 miles from the eastern tip of Cuba to Anaheim, California, where the 1967 major league All-Star game was played. The contest turned into a 15-inning marathon that ultimately was decided by a home run off the bat of Atanasio Rigal Perez, the then 25-year-old third baseman of the Cincinnati Reds.

Victor Perez, setting a record of his own, has no memory of his famous father's clout. The 10-week-old child asleep in his mother's arms was the youngest spectator in the stadium.

The symbolic journey from Camaguey province for Tony Perez, the first of the superstars of the Big Red Machine to depart the Cincinnati scene, was a lot further than 4,000 miles. It was a journey from the sweating, endless toil, back-breaking job of lifting 250-pound sacks of sugar in a dusty refinery. As he worked in the same place his father and two brothers did, two events were taking place on the island that would greatly influence his life.

The less significant one for the country was the fact that Gabe Paul, then running the Cincinnati Reds, was about to corner the market on Cuban baseball through an arrangement with Bobby Maduro, the Havana millionaire owner of the Sugar Kings, then in the International League.

Up in the mountains, the revolution led by bearded Fidel Castro was on the verge of toppling the existing government. That became a reality in January of 1959.

Meanwhile, the Sugar Kings were setting up regional tryout camps for youngsters all over the island. Tony Pacheco, who had been a scout and minor league manager in the Cincinnati organization on the mainland, was checking out prospects in Camaguey. He had a letter from a radio broadcaster who provided him with basic information on a list of 25 names.

"I threw out 18 of the players as being too old or too heavy and had seven names left, one of them being Perez. You know he was never called Tony until he came to the United States. I called him 'Tany' from his name, Atanasio. Anyway, I went to his house and persuaded his father and mother to let me take him to the tryout camp," Pacheco said.

"I liked him and put him on a team that we took around to different places and played maybe 50 games. He did not have a position then. He was skinny, only 160 pounds, but he was very strong. You may find it hard to believe but he could not pull the ball then. His top hand was slipping on the bat. We worked hard to change his grip," continued Pacheco, who came into the big leagues as a coach with the Houston Astros in 1976.

Among Pacheco's students on that tryout team were infielder Chico Ruiz, pitcher Diego Segui and outfielder Tony Gonzalez, all of whom made it to the majors.

The Reds finally signed Tony to a contract and sent him to the Tampa farm camp in March of 1960, one year after the Reds and Los Angeles Dodgers became the last major league teams to play in Cuba. Now, however, with travel restrictions lifted, chances are pre-season games may be played in Havana in 1978.

"Tony Pacheco taught me more baseball than anyonc else," Perez says, paying tribute to his discoverer. "What he did that helped so much was that I played all kinds of positions. He told me the manager in the United States may change me around, and that happened.''

''When I got to Tampa and saw all of those ball players, I was very scared,'' Perez recalls. ''But I did not want to go back to the sugar factory. I hated the noise there. It never stopped. So I told myself, I always dreamed of the big leagues and I must try hard.''

The Cincinnati front office was well aware that patience was a requirement with Latin prospects, who were faced with very difficult adjustments in living in a strange environment. Perez was no exception. He hit only six home runs at Geneva, New York, that first summer and was playing second base when a newcomer from Cincinnati named Pete Rose joined the club. Tony went to third base as Pete took over second.

''It took me a while to get to know Tony,'' Rose said of the man he played with for so long in Cincinnati. ''To tell you the truth, it was hard then to know the Cubans because of the language problem. They pretty much kept to themselves. But even that far back, you could see Tony was different. He had an even disposition.''

The man who eventually got to be known as ''The Big Dog'' or more simply ''Doggie'' in the Cincy clubhouse, was sent back to Geneva a second year. This time he tore the league apart, leading it with a .348 average, 27 homers and 132 runs batted in.

The origins of the ''Doggie'' nickname are hazy. But chances are, it was Dave Bristol, who managed Tony at San Diego and again in Cincinnati, who hung it on him. The implication is that he's the ''big dog'' in the attack.

Another Bristol tribute to Perez later became a private joke in the Cincinnati press box. ''If the game lasts long enough, Tony Perez will win it,'' was Dave's more truth than humor remark as Tony frequently came through in the clutch. Whenever the Reds went into extra innings, there would be a chorus of media voices echoing Bristol's remark — some in sarcasm, some in wishful thinking.

With relations between the United States and Cuba deteriorating in the early 1960's, many Cubans playing ball in this country began to have problems getting back off the island to return the next season.

"I knew when I left in 1963 I wasn't going back," Perez said, despite the fact he had not yet made the majors, was far from financially secure and was only 21 years old. Thereafter Tony established San Juan, Puerto Rico, as his legal residence and generally played in the winter leagues on that island until he became an outstanding star in Cincinnati.

It was in San Juan that Tony met Juana de La Cantera, known to her many friends in baseball and in Cincinnati as Pituka. The close-knit family relationship of Tony, Pituka, and their sons Victor and Eduardo is a beautiful thing, no doubt influenced by Tony's long separation from the rest of his family back in Cuba. It was not until the winter of 1975-76 that he was able to go back to his native island to visit his ailing father. By then he was a hero of the World Series and a slight thaw had set in between the United States and Cuba.

Although Castro always has had a fondness for baseball, it did not keep his government from confiscating all the property and investments of Maduro. When the former millionaire owner of the Sugar Kings arrived in Miami as a refugee, he was without funds. He was given a job in the baseball commissioner's office with a title of representative to Latin-American leagues.

Reams have been written about Tony's even temperament, so much that Perez is occasionally embarrassed about it.

"Cuban players are supposed to be as temperamental as LaScala sopranos, but there is no evidence Tony ever contradicted anyone," wrote Jim Murray, the nationally syndicated Los Angeles columnist in 1974.

"Americans, who get their concept of Latins from Xavier Cugat musicals, tend to think of them as fiery-talking, volatile people who go around dancing with bananas in their hand or throwing shoes or breaking mirrors. But Perez runs more to Gary Cooper than Carmen Miranda," continued Murray.

"I don't ever remember Doggie being mad at anybody on our team," says Rose. "That's a helluva thing to say about a guy after you've been with him a dozen years."

"In the spring you just wind him up and wait for him to drive in 100 runs," says Sparky Anderson, the incumbent boss of the Reds, who managed Tony from 1970 until the Montreal trade in December of 1976.

The amazing thing about Tony's contentment in his years in the Cincinnati organization is that, in a way, he was cheated by fate. Never once was he really the No. 1 man on his team in terms of publicity or fan popularity. When he came up to stay in 1965, Frank Robinson was in his 10th and final year in Cincinnati uniform and Rose was in his third big league season. Vada Pinson was on hand and Deron Johnson led the league in RBI production.

In 1967 manager Bristol did some wholesale juggling by moving Rose to the outfield, shifting Tommy Helms back to second base where Pete had been playing, putting Perez on third base. Despite assuming a defensive burden, this was the year "Doggie" began to come into his own. It was the year he won the All-Star game for the Nationals with the home run that his wife couldn't get to her feet to cheer because she was holding the infant Victor.

The Reds kept Perez at third base to make room for his buddy, Lee May, known in the familiar terminology of the Cincinnati locker room as the Big Bopper from Birmingham.

By 1968 the meteoric Johnny Bench was on the scene and Rose captured the first of two successive batting titles. Eventually Joe Morgan came to Cincinnati and emerged as one of the finest all-around athletes ever to play the game.

If there had been a jealous bone in Tony's body, he would surely have lamented the fate that kept him from being No. 1 when his credentials nearly everywhere else would have been adequate to make such a claim. But in his dozen seasons as a member of the Reds, there was always someone on the scene overshadowing him, which also meant making more money. Perez was the last of the Cincinnati "big four" to cross the $100,000 plateau when that no longer was a measure of ultimate reward in the 1970's.

When it came time to accept the fact a trade would be the best thing to prolong his career, Tony faced up to reality and accepted what was an emotional blow. In being sent to Montreal he was told, in effect, that there no longer was a place for him in the organization he had served so faithfully.

"If someone in the front office told him to walk across the Ohio River, he'd have tried it," wrote veteran Cincinnati baseball writer Earl Lawson.

The raw power of Tony's swing was apparent from the beginning. His arms and shoulders showed the development of moving those dusty bags of sugar in the heat of the Cuban sun. But the other side of Tony's success as a hitter was relatively slow in asserting itself.

One of his 12 home runs in his rookie season was one of the longest ever hit in Milwaukee's County Stadium, now the home of the American League Brewers. Tony Cloninger, the pitcher who came to the Reds in a 1968 deal from the Braves, remembers it well.

"The ball cleared the left field bleachers and was still going," Cloninger said of the blast. "I'm just glad it was Dan Osinski who threw it, not me."

But for at least three more seasons the standout righthanders in the league had two ways of handling Perez. The cagey ones, such as veteran Jim Bunning, and the tricky Juan Marichal disposed of Tony by having him chase pitches over the low and outside corner of the plate. The fire-balling Bob Gibson, then in his prime, did it with the high hard "jammer" — but he was getting nearly everyone at that stage.

Every winter during the late 1960's Perez played in the Puerto Rican winter league, partly for economic reasons and partly as a laboratory to sharpen his hitting techniques. In the spring of 1968, he told of what he had done in winter ball to sharpen his tools.

"When I faced a hard-throwing righthander, I think it is Bob Gibson," he explained. "That may not help me against Gibson, because nobody throws like he does. But against breaking ball pitchers I would swing at a lot of outside, low pitches that maybe I shouldn't have just to make contact on the pitch out there."

His spiraling batting average was evidence of the success of his concentration as it climbed from .282 to .294 to .317 in the three seasons of 1968-69-70. It was Tony's white-hot bat that helped the Reds get off to the sizzling start that saw them win 70 of their first 100 games in Sparky Anderson's first season and make a joke of the division race.

In April of that year Tony hammered 10 home runs, drove in 26 runs and batted .454. In Tony's words his Puerto Rican laboratory on swinging at breaking pitches had paid the dividend he wanted.

"I now can judge whether the outside pitch is in the strike zone or not and I can lay off pitches I used to swing at," is the way he put it.

Larry Shepard, then the new Cincinnati pitching coach, summed up Tony's ability. "He's reached the stage where only the good hitters get," said Shep, who has managed some good ones including Roberto Clemente at Pittsburgh. "There's no sure way to get him out. There is no set formula. You have to mix things up and be careful not to make a mistake."

Family man that he is, Tony had added motivation at this point in that second son, Eduardo, had arrived to increase the family size to four. Both Perez youngsters are showing a tendency to grow into sizeable teenagers, which is hardly surprising in that both parents are tall. When Victor was only two years old, Tony said of him, "He is a big bear, always doing something, never stops."

In one sense, it seems odd that the amiable slugger has never again come close to the 40 home runs he hit in 1970 — except to remember that was the year the club moved out of "Cozy Crosley" into Riverfront. Both Perez and the much younger Bench had to make some adjustments in their batting style in the more spacious dimensions of the new playing field, and this was reflected in the 1971 downfall of the club.

As Bob Howsam revised the caliber of the Reds with the spectacular deal with the Houston Astros involving Joe Morgan, not much attention was paid to the role Perez played in the deal. Slugging Lee May was the "bait" the Astros got hooked on. But the Reds couldn't have given up a slugger of May's caliber without Perez to shift to Lee's position and help pick up the slack in the batting order.

The statistics of Tony's regular season contribution to the Reds' pennant drive in 1972 are not spectacular by his career standards. But in the disappointing World Series against the Oakland A's, Tony's .435 average played a prominent role in carrying the struggle to seven games after the A's won the first two in Riverfront.

By this stage in his career, Perez was known as one of the best "two strike" hitters in the game.

"You don't worry about him with two strikes," manager Anderson puts it.

Perez is the only player active who has driven in 90 or more runs over 10 consecutive seasons, a not inconsequential statistic that such veteran RBI producers as Willie Stargell, Willie McCovey and Carl Yastrzemski have not duplicated.

"I am a better hitter with men on base," Tony says with candor. "I have been asked why many times, and it is hard to say. I do concentrate. It is a challenge. I like to walk up knowing we need a hit."

Tony's temperament is an asset in this respect, of course.

"He blows hot and cold," says Ted Kluszewski, the able Cincinnati batting coach, "but even when he's not swinging the bat well, he has the knack of knocking in runs."

On the night of September 2, 1975, in Riverfront Stadium, Perez drilled a run-scoring single off San Diego's Jerry Johnson for the 1,010 RBI of his career, breaking a Cincinnati club record that had been owned by Frank Robinson. As the crowd stood and cheered and the umpire handed him the ball, Tony in turn gave it to the bat boy with instructions to take it over to the railing near the dugout where Victor and Eduardo were cheering.

"I am very proud to have taken a record of Frank Robinson's," he said afterwards. "He has been a great player in both leagues."

In Game Two of the 1975 championship series, Perez ripped a two-run home run in the first inning to give Fred Norman a working cushion toward a 6-1 victory.

But the most meaningful and most dramatic of the Perez home runs was to be delivered in Boston's Fenway Park during that pulsating 1975 World Series, which began on a "down" note for Big Tony who had no hits in the first four games. His 0-for-15 performance was approaching in futility a streak of the late Gil Hodges, who went hitless in 21 times against the New York Yankees in the 1955 Series. That development led to a Brooklyn priest publicly urging his congregation in a Sunday sermon to "pray for Gil" that afternoon.

There is no record of divine intercession for Perez, but in the fifth game he hammered a fourth inning home to tie the game at 1-1. He added another blast in his next trip to help the Reds win the game and move within one more victory of that long-awaited World Championship.

In the seventh and final game the night of October 22, the Red Sox scored three times in the second inning, capitalizing on Don Gullett's wildness, as Bill Lee, an eccentric southpaw, was pitching his second strong game of the Series. There were two men out with a man on in the sixth, with Boston still leading 3-0, when Perez stepped in. Earlier Lee had made him look bad on a slow, blooper-type of a change-up.

Tempting fate, Lee tried it again and this time Doggie wasn't taken by surprise. He put the ball into orbit, sending it soaring into the night sky far over the left field screen. "When last seen, the ball was headed toward Daniel Webster's grave up in New Hampshire," wrote one excited Ohio columnist.

Brought within one run by their clutch specialist, the Reds then tied the game on a single by Rose and won it on a base hit by Morgan.

There were champagne-soaked tears in the clubhouse and no one was more emotionally involved than Perez. Tony had been hearing and reading of trade rumors, but at least if he had to go he had helped carry the Machine to the top of the mountain in his 11th season.

But there was to be a 12th before Tony's era with the Machine was terminated and, once more, he came through. It seemed for a while that Tony's try for a 10th year of 90 or more RBI was doomed to fail but a late surge carried him to 91.

In the playoff against the Phillies, he drove in three runs and in the World Series sweep over the Yankees, Tony drove in a run and had three hits in the 5-1 opener. His single off Catfish Hunter broke up the 4-3 second game in the ninth inning.

Finally, in December of 1976, with two first basemen in the wings, including Dan Driessen who had been with the club since 1973, the Reds did trade Tony to the Montreal Expos. His approval was needed, and he accepted with the Canadian team offering him a three-year contract at $200,000 per.

"It is an opportunity to play and the money is good," Tony said dutifully at the announcement in Cincinnati.

How much will the Reds miss Perez the person and Perez the clutch hitter?

Dodger catcher Steve Yeager, representing the point of view of the enemy, believes it will be plenty. "Of all the guys on the Reds, and there are plenty of tough hitters, the one I hated to see walk up there with the game on the line was Tony. To me, he was the toughest out in the clutch, tougher than Rose, tougher than Morgan, tougher than Bench," Yeager put it.

As if to emphasize that point, Perez hit a home run in his 1977 league debut for the Expos in Philadelphia, hit a second one the next day and was a popular figure in his new surroundings even before making his debut in Canada in the stadium built for the 1976 Olympic Games.

During spring training, Tony, in his powder blue Montreal uniform, conceded that "it is like moving to a new family" with the Expos. He admitted that Pituka had telephoned Karolyn Rose several times in lonely moments in her new surroundings.

There is no doubt in the minds of the people who knew and admired them so much in Cincinnati, that Tony and Pituka will cast a meaningful and probably delightful shadow over their new environment in Canada.

Years ago, back in their tiny house in Camaguey, Tony had told his mother, "Some day, I'm going to be a ball player as good as Minnie Minoso." Taking nothing away from the colorful Minoso, who rejoined the Chicago White Sox as a coach in 1976, Perez has outdistanced him.

But the era of The Big Dog in Cincinnati is over. And if it seems curious that a man of Tony's class and loyalty is remembered by a seemingly unflattering nickname, remember that no man was ever verbally abused with more respect and genuine affection from the people who used it.

THE COBRA

It never occurred to Dan Driessen as he was growing up in Hilton Head, the South Carolina resort community just north of Savannah, Georgia, that he could be classified by sociologists in just about every known category of the "disadvantaged." Raw-boned Dan is black, he was poor and, as he discovered later, considerably undereducated. His father died when he was six and he's one of eight children.

But the man who replaced Tony Perez after a four-season apprenticeship with the Cincinnati Reds isn't one to lament what never came his way.

"We didn't think we were all that poor and it never occurred to us that the day would come when we could go to school with the white kids," Driessen says. "And I wasn't bussed until I went to high school. In fact, we lived only about twice as far from grade school as the distance from home plate to centerfield in Riverfront."

Driessen's mother worked in one of the larger resort hotels in Hilton Head. "And I waited table and was a bellman in the same hotel while I was going to school," he says.

Everything seemed to come along a shade late in the years Danny wa growing up.

"They started a football team my junior year in high school and I had no interest in playing," he says. "The year after I graduated, they integrated the schools."

Part of the reason Driessen, who certainly had the physical potential to play football, disdained that sport was that he was so deeply involved in baseball. In the segregated high school at Hardeeville, South Carolina, the baseball team, coached by one Harold J. Young, had a brother battery. Bill Driessen was the pitching star and his catcher, by default, was his little brother, Dan.

Schoolteacher Young eventually wrote a letter about the Driessen brothers, with the emphasis on Bill, had it mimeographed and mailed to clubs around the majors. He described Bill Driessen as 6-foot-3 and 215 pounds, all muscles and shoulders, with good speed and range and with the ability to play virtually every position on the field.

"He exaggerated by an inch or two and a few pounds," laughs Dan. "He said I was a 190-pounder in that letter, but I never weighed close to that, even though people called me 'Tubby' when I was in high school."

It is a policy of the Reds under Bob Howsam that any such letters of recommendation are not thrown into the waste basket. Teacher Young's letter was referred to Bill Jamieson, the primary scout in the Carolinas. The Driessen brothers were invited to come to nearby Savannah to work out with the Asheville, North Carolina club, at the time a Cincinnati farm team. This was in the summer of 1969 and manager Alex Cosmidis was instructed to file a report on the pair.

Cosmidis wasn't all that impressed. His letter back to Jamieson was negative on Bill's prospects and lukewarm on Danny, but he did say the latter swung a bat pretty well. Jamieson then paid a visit to Hilton Head and had his own tryout for the pair.

"Bill wasn't the sort of pitching prospect we were looking for," Jamieson said. "But I liked what I saw of Dan — a lot of speed and a good stroke."

So Driessen was signed, as he says now, "for a yearbook and an airplane ticket to Tampa.'' When it came time to go, he missed the flight and went down on the Greyhound bus, arriving a day late.

''Funny thing was, the bus had a flat tire along the way and I got off and helped the driver change it,'' Driessen recalls, ''and he wound up giving me $5. He didn't know how worried I was about being late to camp.''

Dan was listed as a catcher. Ron Plaza, one of the seldom publicized but very valuable instructors in the Cincinnati farm system and a onetime catcher himself, did not like what he saw of Danny's movements behind the plate. Plaza and the other instructors also were puzzled by Dan's lack of communication. He seldom spoke and when he did it was such a low-keyed voice that he was extremely difficult to understand.

''There was no way this kid could communicate with pitchers, so we decided he'd be a first baseman,'' Plaza said later.

His first season in pro ball wasn't easy. He was homesick and he didn't get along very well with his manager. But a year later Russ Nixon was managing the Tarpons and he built a very strong relationship with Driessen, who was more self-confident by then and better adjusted to the rigors of life in the low minors. He batted a robust .327.

To his great surprise, he was invited to spring training with the parent club in 1972 because Sparky Anderson wanted a look at his quick bat, having been impressed by Danny in the Instructional League the previous fall. Despite having six hits in nine at bats, he was farmed out to Three Rivers where he had a delightful year, batting .322 and hitting 37 doubles.

''Everything was nice up there but the weather,'' Driessen said of his stay in Quebec province.

Sparky had nicknamed him "The Cobra" for his swing, saying that "he uncoils like a snake. He's a born hitter."

Looking ahead going into 1973, the plan was to try to convert Driessen into a third baseman because Tony Perez was in his prime at first base with the parent Reds. On the other hand, Denis Menke, who had played third in '72, had hit only .233 and was showing definite signs of slowing down even though he was nearing only his 33rd birthday.

Sparky hoped to go with Menke for one more year and give Driessen time in Indianapolis to familiarize himself with the new position. But Menke's hitting was so anemic that a crisis developed early in the season while Driessen was batting at an incredible .409 pace in 47 games in the Hoosier city.

Danny was stunned when he reported to the Reds in Chicago and found himself in the lineup. He was never out of it despite the fact that Chicago's Fergie Jenkins struck him out — with the bases loaded — in his first major league at bat. At third base Driessen did exceptionally well considering his limited experience at the position. There was virtually no criticism of his play until a mental lapse in the championship series with the Mets. His .301 batting average helped pep up the Cincinnati offense.

The stormy showdown between the Mets and the Reds, already punctuated by Pete Rose's wrestling match with shortstop Bud Harrelson, moved into a fifth and decisive game. Jack Billingham, the Cincinnati starter, gave up two runs in the first inning but the Reds came back to tie it 2-2 with runs in the third and the fifth. Wayne Garrett led off the bottom of the fifth with a double and Felix Millan was given the sign to sacrifice him to third. But Millan's bunt was too hard, straight back at Billingham who threw to third. Driessen took the throw but, to the consternation of everyone in the Cincinnati dugout, did not apply a tag to the sliding runner. Before the inning was over the Mets had scored four runs.

Driessen, sitting with his head in his hands in the Cincinnati clubhouse, did not attempt to alibi.

"I thought there were runners on first and second and I had a force play," he said softly. "I blew it, pure and simple."

Whatever chance Driessen had to become an adequate third baseman, as Perez had done before him back in 1967, went down the drain with that mental lapse.

"From then on, all I heard was what a bad third baseman I was," Driessen says in looking back. "I went to Puerto Rico that winter to keep working at the position, but I might as well have stayed home."

Midway in the 1974 season, Anderson came to the conclusion that Driessen would be an outfielder.

"We put an awful lot of pressure on Danny by using him at third base, but that play in New York turned everything around for him. He couldn't cope with the pressure," the manager said. The problem, however, was that the Reds didn't have anyone else. Darrel Chaney saw a lot of service there and outfielder Merv Rettenmund even started two games at the position.

So Driessen trotted off to Puerto Rico once more to play winter ball and to try to adjust to playing the outfield. The process was interrupted when he suffered a hairline fracture in his left wrist — playing first base.

The outfield Driessen was trying to crack in spring training of 1975 had Pete Rose in left, Cesar Geronimo in center and the still unproven Ken Griffey in right, not to mention George Foster on hand.

With the club spinning its wheels in the early going,

Anderson made his daring move of switching Rose to third base. At the time, he talked of a platoon opening in the outfield for Foster and Driessen. But the former, getting the first opportunity, took care of that by establishing himself as one of the more potent hitters in the league.

Driessen suddenly found his playing time dropping off dramatically — from 150 games to 88 and 410 to 210 at bats. Amazingly, for those 150 games his .281 batting average was identical to his 1974 figure.

"I realized then I wasn't going to be making the kind of money I'd like sitting in the dugout," Driessen put it. "But, of course, when we got into the World Series that helped take some of the sting out of it."

What was bothering Danny, and bothering the Reds as well, was the possibility that his natural hitting talents could "rust out" on the bench.

"We've got to play Danny a little more," Anderson said in spring training of 1976.

"People keep telling me my time will come," Driessen lamented then. "You sit on the bench and you lose your stroke. I've been a pretty good pinch hitter but it's not the same as playing."

Somehow in the heat of the race Sparky never got around to finding more playing opportunities for his "Cobra." Danny's average tailed off to .247, although the fact that he drove in 47 runs proves it wasn't completely a "lost" effort for the easy-going South Carolinian.

Driessen had the satisfaction of becoming the National League's first designated hitter, playing all four games in the World Series sweep over the Yankees. Starting slowly, he wound up with a .357 average, hitting a home run and a double to help Pat Zachry win Game Three.

Danny had a talk with Bob Howsam in September,

asking the club to either assure him of an opportunity to play or to make a trade in the off-season. Given that option, Howsam concentrated on an attempt to make a reasonable deal for veteran Perez, realizing the eight-year difference in age between the two first baseman dictated that the older one should go.

To his amazement, Driessen found that he was still battling Perez in a sense during the 1977 spring training period.

So the young man went into 1977 carrying a mental burden that he didn't suspect would be there once Perez had finally left the position to him. The surprising fact is that Driessen had signed a one-year contract before the Reds announced the trade of Perez.

"They made me what I considered a very good offer," he explained. "In fact, it was enough to indicate I was getting a regular's salary. They also indicated we could talk about two years but I didn't want to get into that. I'd like to think my value and the bargaining power will be better next winter."

Driessen was married shortly before going to spring training in 1976. His wife, Bonnie, gave birth to their first child, a daughter named Dominique, while Danny was again playing winter ball in Puerto Rico in late December of '76.

"So I got more things to think about and more reason to play good now," he said, grinning.

"Danny has real good speed for a good-sized man," Manager Anderson says. "He'll get some leg hits and once he really settles down he should be a good, steady hitter. I thought he could do that the first time I ever saw him."

Maybe so, but it took the young fellow from Hilton Head a long time to get the green light for a try at doing it.

Danny's strong follow through

The serious side of Billingham

EASY RIDER

Lanky Jack Billingham has stretched his 6-foot-4 frame to its full length, reached out with a firm grip and just about pulled himself atop the pitching mountain. But always, as he maneuvers for the final effort to reach the pinnacle, misfortune in some form seems to obstruct the friendly Floridian, who deserves a kinder fate and more recognition for his contribution to the Cincinnati Reds.

Billingham may need a public relations man in one sense, but it's hard to talk to him and not come away with respect for a likeable human being. There is a whimsical strain, not often found in athletes, that enables Jack to take his success in stride and accept his difficult moments without complaint.

He achieved his 100th career victory on a windy, raw day in Chicago's Wrigley Field on May 7, 1976. The wind was blowing in with Arctic gale force off Lake Michigan and only a handful of well-bundled spectators were moving around the park as he pitched a crisp 3-1 victory. Afterwards in the warmth of the clubhouse the media approached the big blond for his comments on the milestone, not uncommon but still meaningful in the life of any pitcher.

"It's nice," Billingham said, "but as a milestone, well a lot of people have done it. I just wanted to get it over with and get everybody in out of the cold."

During the 1974 season when his shoulder was stiffening on him at some stage of every game he started, Jack hardly missed a turn.

"He pitched a whole season in pain and never complained once," manager Sparky Anderson said of him in September.

"You don't rate a hero's badge for throwing a baseball," Billingham said when the compliment was relayed to him.

When Jack's celebrity teammate Johnny Bench was wed to Vickie Chesser, the list of those invited was headed by President Ford. The rest of the guest list sounded like a Las Vegas spectacular. Police had to set up crowd and traffic control posts around the church. Television cameras were spotted between banks of flowers at the altar.

Paul Meyer, the baseball writer of the Dayton Journal Herald with a flair for the whimisical approach himself, asked Jack and Jolene Billingham about their wedding.

"It wasn't even the social event of the season in Winter Park, Florida, let alone the whole country," the pitcher said. "There wasn't a single company president there and it didn't slow down traffic by the church," Jolene added.

Neither the Billinghams nor writer Meyer was making fun of the Bench soiree. It was a comparative thing about two ordinary, quiet people — maybe the "Ordinary Couple" from "The Sound of Music."

Jack's easy-going temperament, which isn't likely to change, may have held him back in the early years of his baseball career with the Los Angeles Dodgers.

"Somebody was always trying to fire me up," he said, looking back. "But it's not my nature. Tom Lasorda (who replaced the legendary Walter Alston as manager of the Dodgers going into the 1977 season) was a minor league pitching coach and a real gung-ho guy.

"He was always yelling things like 'you gotta love it,'

'get mean,' 'knock somebody down,' 'you're gonna bleed Dodger blue' and things like that. I always thought he was kind of funny, not that he isn't a nice guy, but I can't accept basball as a life or death thing."

Billingham signed into the Dodger organization out of high school in 1961 and it took him a long seven seasons of minor league stops to reach the majors. He was groomed as a relief pitcher. His lengthy apprenticeship has given him a sense of appreciation of baseball hardly typical of the general dollar-sign attitude and conversation of the present era.

"When I was a kid growing up, my father owned a gasoline station and I spent a lot of hours working around the place when I was old enough to help out. We didn't have a lot, but we lived comfortably. When you don't have a lot of things, you don't miss them.

"Now that I'm making more money than I ever thought I would, we have a nice car and a nice home in Florida with a swimming pool. I never thought about a swimming pool as a kid. I'm appreciative of everything my father gave me and I think I have a deeper appreciation of what baseball has given me than a lot of younger players who spent only a year or two coming up."

Billingham always has maintained that his easy-going mannerisms don't mean that he isn't a good competitor when he's pitching.

"I want to win as much as the next guy," he says, "and I concentrate on what I'm doing. I'll admit that once I had a tendency to let down against second division teams or light hitters. I think I've learned my lesson in that. I get a lot of mail from fans telling me to get fired up. But the Reds don't mess with me. They take me for what I am."

What Billingham has been to the Reds in the five seasons since accompanying Joe Morgan from Houston is best appreciated by comparing his statistics with those of Don Gullett, the young lefthander who played out his option to sign a $2.1 million free agent contract with the New York Yankees after the 1976 season.

In the five seasons (1972-1976) they were together, Billingham won seven more games than the young Kentuckian and, more impressively, started 167 times to 123 for Gullett. Even conceding the latter's time lost to injuries, Tall Jack has been every inch as much a workhorse as his better publicized friend.

Billingham spent his first season in the big leagues in the Dodger bullpen, compiling a 3-0 record in 50 appearances, all but one in relief. His name was on the available list for the expansion draft when Montreal and San Diego came into the league prior to the 1969 season. The Expos claimed him but he never pitched a game for the Canadian team. He was traded to Houston a week before the regular season was to open and plunged into a soul-searching crisis that had nothing to do with baseball.

The Billinghams had a two-year-old son at the time but Jolene was pregnant and having some problems with it when the Montreal-to-Houston deal was made.

"It caught us by surprise, but we agreed to go to Houston. We lived for two weeks in the Astroworld hotel across from the Astrodome until we found an apartment. The day we moved into the apartment, Jolene started hemorrhaging, and she was in and out of the hospital for a month," Jack said.

The doctors were struggling to prevent a miscarriage and it was a touch-and-go struggle.

"Finally, the doctors said the baby might have a chance to survive and it would be best to let it happen. Jolene was getting torn up mentally and physically too," her husband recalled.

Jennifer Billingham came into the world in a precarious and tiny state, being destined to spend 76 days in a Houston hospital. But the ending is a happy one and today she is a healthy, happy youngster.

"It was a real jolt for me to have to go on the road and leave Jolene," Jack said. "She had our little guy and wasn't all that strong herself. They later told us it was a miracle that Jennifer survived. But after the ordeal, we're a very close family with two fine youngsters."

Jack struggled through a mediocre year in the Astro bullpen, but there was a turning point from it. The next spring the Astros gave him a chance to move into the starting rotation and he made the adjustment quickly.

"I'm a control pitcher and it's easier to work when you have a full warmup to get things together. I never rebelled temperamentally against pitching in relief, but I felt I'd be much better as a starter."

He had a 13-9 season, but then slipped back to a 10-16 record in 1971 just before Reds' President Bob Howsam pulled off a trade that will always be remembered as the "Joe Morgan steal." But the Billingham dividend has been an impressive one.

"I came to the Reds with the reputation of being a second half pitcher," Jack laughs, his whimsical tendency surfacing easily. "I lost my first five decisions."

But if Jack struggled a bit to attain his 12-12 record as the Reds won the pennant, it was a worthwhile season in that he earned the confidence of his new manager, Sparky Anderson, and pitching coach, Larry Shepard.

His finest hour came in the World Series against the Oakland A's, who had taken the first two games in Cincinnati. As he prepared to open the three-game set in the California city, no one was more aware than he that it was up to him to halt the stampede. But before the first game could get underway, a freak downpour innundated Oakland Stadium during the evening rush hour, forcing postponement of the contest and drenching thousands of persons around the Stadium.

"I had already had two days to think about my first World Series game," Billingham recalled. "Now it was getting out of uniform, going back to the hotel with an evening to spend and another night to sleep on it. But we had a nice dinner with the Darrel Chaneys and I had a good night's sleep."

Whatever the reason, he came out and pitched a masterpiece against Charles O. Finley's green-and-gold athletes and set the trend for what turned into a seven game struggle. Clay Carroll came on in the eighth to finish Jack's 1-0 victory. Eventually Jack pitched 14 innings against the A's and gave up only one unearned run.

The next spring Anderson touted Billingham as the new workhorse of the Cincinnat staff and Jack responded to the challenge very well. He finished with a 19-10 record with a 3.04 earned run figure. The 293 innings he pitched were the most by a Cincinnati pitcher in 32 years. However, the season ended on a controversial note and an unexpected second guess of his manager by Billingham.

Bidding for his 20th victory, Jack started a meaningless season's finale against the San Francisco Giants. He didn't have real good "stuff" and the Giants piled up a 4-1 lead with the Reds coming to bat in the sixth. Manager Anderson lifted his pitcher for a pinch hitter. The Reds never caught up but Billingham later had this complaint:

"I think I had earned the chance to stay in there in case we did rally. The game meant nothing."

His World Series effort in 1972 coupled with the 19-10 indicated that Jack was on the verge of moving front and center among the league's established pitching stars.

"The World Series did a lot for my confidence in myself," Jack admits. "If nothing else it stopped people from saying 'Jack Who' when they heard my name. The next season everything came together in my mind and I was confident I was on the verge of really establishing myself."

But fate was about to deal him a blow in the form of a shoulder problem that turned the 1974 season into a nightmare of struggle and pain. In the records, Billingham's 19-11 effort was somewhat identical to his successful surge the year before.

"But it was a much different year," he explained. "I was taking cortisone by the quart. Once the club flew me home from Pittsburgh for a visit to the doctor and got me back before anyone knew I was gone for the day. I'd start off and might be pitching well, then the shoulder tightened, and I'd lose my rhythm and velocity. It's incredible, but I had four shots at winning my 20th game in September and couldn't make it. It was clearly my fault then."

Chances are, Jack shouldn't even have been pitching in September but the Reds were still chasing the Dodgers and he stayed in the rotation.

The winter of 1974-75 was a difficult one. Always the realist, Billingham knew that if his shoulder problem was chronic he was at an age where no club is apt to stick with a player very long. Although he was advised to give his pitching arm complete rest, he went on a rugged conditioning program.

Even though 1975 turned into the long-awaited year the Reds won a World Series, it was up and down for big Jack. His shoulder problem was fairly well healed, but for the first time in his career he had long stretches where his control deserted him.

"Shep and I figured out that I had gotten into some

bad habits when I was pitching with the bad shoulder,'' he said. ''Finding them is one thing, correcting them another. I couldn't make the pitches I wanted.''

With a third championship series coming up with Pittsburgh in six years, Sparky announced a pitching rotation that had lefty Fred Norman along with Gullett facing the Pirates, who displayed a heavy emphasis on left-handed hitting power.

''I was hurt when I learned I wasn't going to be in the rotation,'' Billingham admits. ''But I couldn't fault Sparky. I wasn't pitching well and Norman was a logical choice. But I couldn't help but feel maybe I was losing my last chance to help and be a part of a championship team that went all the way.''

He did sit out the sweep over the Pirates but Anderson restored him to duty for the World Series with the Red Sox. The Boston team had a batting order with a right-handed emphasis plus the wall at close range in left field. Billingham started the second game and held the Red Sox in check until the sixth inning when the bullpen wheeled into action and the Reds evened the Series.

Jack was scheduled to start the sixth game when the Series moved back to Boston. This time three days of rain washed out Jack's return but he did pitch two scoreless innings in relief in the seventh and final game when the Reds fought back for a 4-3 win and the championship.

''What Jack did in the Series made his struggle over the last half of the season amount to something,'' said Shepard in admiration of a hard worker.

After the Series, Billingham revealed that he had been undergoing hypnotic treatment. The first suggestion had come from Houston pitcher Larry Dierker, an ex-teammate who had gone that route himself. Bob Howsam surprised the pitcher with a similar suggestion.

"I went to the guy and he relaxed the hell out of me," Jack says in the form of a recommendation.

Despite the hypnotist and on his finger the World Championship ring with a diamond embedded in the red ruby, Billingham went through another roller coaster experience during the 1976 championship season. Demoted to the bullpen at one stretch, he had only a 12-10 record to show for his efforts. Still there were some impressive outings, including an early shutout over the Dodgers and his previously mentioned 100th career victory.

But once again he had something in reserve for the World Series, gaining credit for the 4-3 second game victory with three scoreless innings in relief. Catfish Hunter was the loser.

Going into the 1977 season at age 34, Easy Rider was as philosophical as ever.

"I've reached the stage where I know I can't go on forever. I'm going to be taking a summer vacation with the family one of these years. That's something a ballplayer can't enjoy until he's through.

"But maybe, just maybe, that 20-game season is floating around somewhere out there waiting for me to pick it off," he said. "For the last two seasons, my biggest problem has been in my mind. I'm thinking too much and worrying about how I'm going to do things instead of merely doing them.

"I think this is a big part of baseball and maybe life. Whenever anyone, pitcher or hitter, is in a slump, they mess themselves up by thinking too much. Yet it's a natural thing to do. Before I went to the hypnotist, my mind was playing tricks on me, night and day."

One thing that isn't likely to change is Jack's temperament. His "dark side" ledger is thin.

"I've only thrown at one batter to hit him on my

own,'' he said. ''Tony Gonzalez laughed at me one day after he fell down missing one of my sweeping slow curves. In my mind he was making fun of me. I said to myself. 'You're going down this time and you won't be laughing'.''

Strangely Jack is on record as triggering the only real gang fight the Reds have had in the 1970's, a donnybrook in Pittsburgh in 1975. But the real villain was Pirate pitcher Bruce Kison who was ''shaving'' Cincinnati batters with inside fast balls.

''When Kison came to bat, I had to hit him, and you know that wasn't strictly my idea,'' he claims. ''It was a real good brawl and I managed to sort of stay around the fringe of the real action. It's too bad though that Tom Lasorda couldn't have been there.''

That's Easy Rider — always whimsical.

But there's nothing whimsical about his pitching contributions to the success of the Reds.

THE QUIET MAN

For years Cesar Geronimo had enjoyed this day-dreaming fantasy of playing in Yankee Stadium and at long last it was coming true. But on October 15, 1976, when the slender, silent ballhawk first set his spikes in the hallowed turf of the then-remodeled "House That Babe Ruth Built" he was not wearing the traditional pinstripes of the old Yankees.

"My dream as a boy always had been to play alongside Mickey Mantle," explained Cesar. "But the Yankees as a team were my heroes. They always had been and I knew of Babe Ruth and Lou Gehrig and Red Ruffing and all the old heroes.

"But at this minute," he went on, 'I would not have wanted to trade uniforms."

The uniform Cesar was wearing belonged to the Cincinnati Reds where he had become a Golden Glove centerfielder and played a role as the Machine swept the proud Yankees in four straight games for their 1976 Series triumph.

Cesar's first visit was when the Reds went for their workout the day before Game Three action resumed. He admits there was a lump in his throat as he emerged from the tunnel leading from the visitor's locker to the dugout.

"I looked around and of course it is not now what I used to know in pictures," he said, "but it was Yankee Stadium and it was hard for me to believe I was there."

Oddly, the Yankees once had title to Cesar Geronimo's contract but permitted it to slip away to the

Houston Astros. The truth is that in his two seasons far down in the Yankee farm system, Cesar had difficulty hitting his weight, which isn't all that much.

His love for the Yankees dates back to his early boyhood in the Dominican Republic. In truth, it did not appear that Cesar was destined to become a baseball player. At age 12, he was enrolled in the Santo Tomas De Aquino Seminary to study for the priesthood.

"It was run by the Jesuits, but I was not going into the Jesuit order," he said of the Roman Catholic religious order noted for its scholastic excellence and lengthy preparation of its priests.

Cesar believes his early inclination to become a priest was legitimate enough, but that he really was too interested in sports to balance the two influences. At night when the good Fathers gave the "lights out" order to their youthful seminarians, Cesar would dutifully crawl into his bed. But instead of going to sleep, he listened to baseball broadcasts on a small radio he would sneak under his pillow.

"I listened to any games that I could get, but we got a lot of Yankee games and when we did that made me very happy," he said, referring to the Spanish-language radio broadcasts out of this country, the majority of which originated in New York City.

After four years the youngster came to the conclusion that the life of a priest was not for him and withdrew from the seminary with his family's consent. Perhaps it was here that his life-long habit of reserve was deeply embedded into his character.

"I liked it at the seminary," he recalled, "but I really wanted to be a ball player and I knew if I went on to become a priest that would never happen."

Ironically, despite his dreams of far-away Yankee

Stadium and the legendary greats who played there, Geronimo had no baseball experience.

"At the seminary," he explained, "the sports we played were basketball, soccer and softball. I was a very good player in soccer and the other boys called me 'Super Player' which was nice, but I never got enough of them interested in playing baseball."

Geronimo's most obvious asset in baseball, and the one that lured the professional contract for him, was his throwing arm, described by Reds manager Sparky Anderson as "my cannon in centerfield."

"I was playing softball on the same team with my father," Cesar recalls. "It was he who took me to tryout camps. First, we went to the Mets because they had one in El Seibo where we lived.

"Then we went to a Yankee tryout camp run by Pepe Sada in another city, and they did not offer to sign me either. But they did invite me back and signed me then. The first time they were looking for pitchers."

The fact is that when Cesar was sent to the mainland he was listed as a pitcher-outfielder and he does not know why the Yankees did not attempt to use him as a pitcher in the minors, especially in view of his horrible batting statistics. "I did not like to pitch, and I am happy they didn't," Cesar says.

Geronimo's baseball swing was incredibly bad because of his experience at swinging at slow pitch softballs. When he came to the Reds, batting coach Ted Kluszewski says his swing was "absolutely as bad as anything I'd ever seen in a big league uniform. He uppercut everything he swung at and hit some of the softest pop flies you ever saw."

Even Anderson, who now calls him the "best centerfielder in baseball," never believed he would turn into a consistent hitter.

"I give Teddie (Kluszewski) all the credit in the world," Sparky says. "I never thought Cesar would hit but I couldn't say that publicly when he joined us. I knew we had a big league glove man but I honestly despaired of him ever having a chance to hit."

In retrospect, it seems odd that the Houston team did not do a better job in restyling his swing. Even after trading him in the famed Joe Morgan deal, Spec Richardson, then the General Manager of the Astros, talked highly of Cesar's potential.

Geronimo spent the entire 1969 season with the Astros and went to bat only eight times, but he appeared in 28 games being used mostly as a defensive replacement. He was with the Columbus, Georgia, farm club half of the 1971 season and, adding it up, the softball swinger had only 612 at bats in his minor league career.

But Geronimo concedes it may have been more than his weak swing that somewhat retarded his progress. He is so reticent and seemingly uncommunicative that he perhaps discouraged attempts of instructors who lacked Klu's patience to work with him.

"We had a problem in that regard," Kluszewski admits. "And it was very hard for me to know if I really was getting through to him or not. But for that reason, I kept prodding him about the one thing he had to learn to do before anything else. That was to get on top of the ball, hit it on the ground. When he made real good contact he'd still get line drives. But balls on the ground go through for hits a lot of times."

Geronimo had another problem to cope with before coming into his own after the Reds provided him with an opportunity. He isn't as physically strong as the average big league outfielder. Making that sort of admission never is easy for a professional athlete but Cesar has come to terms with the problem. He has cut back on his winter baseball as an energy conservation matter for his major league season.

"I realized going into the 1976 season that no matter what else I would do to build up my strength, I must be careful during the hot months, so I skip a lot of batting practice," he says.

Skipper Anderson augments that with an occasional day off, and Geronimo no longer resents that happening, although it took late season batting slumps in both 1974 and 1975 to convince him.

"In both of those years I felt I would finish hitting close to .300," he says. But the fact is he tailed off to .281 in 1974 and then had an even greater late season drop in efficiency in the championship campaign of 1975.

When he hit .307 the next year, Cesar felt he had discovered the formula for conserving his physical assets and bringing out the best in his potential. "I would still like to have a better year," he said going into the 1977 season. "I would like to hit around .320."

Geronimo's above-average speed helps him get his share of infield hits, but he never has developed the art of bunting, partly because it was such a lengthy struggle to get him to the point of controlling his stroke to eliminate all the early faults.

It is ironic that Geronimo, whose career high for home runs in a season is seven, hit two in the 1975 World Series. When he and Dave Concepcion delivered back-to-back shots off Boston's Rick Wise in Game Five it marked the first time any National League duo had achieved that in Series play.

"He (Wise) was throwing me nothing but fast balls. I had fouled off three of them and I really was taking my best cut on the one I hit out, but no, I did not think I could hit a home run. I never think that."

Although he has a volunteer press agent in Manager Anderson, Geronimo is surely the least known of the Cincinnati regulars from the 1975 and 1976 championship teams — and he admits he is a little bit bothered by that even though he understands it.

"It would be nice to be better known," he puts it. "But I know why Joe Morgan, Johnny Bench and the others get more attention and I do not blame people for wanting to talk to them and not me. As long as I do my job and am paid well for it, I will not be too unhappy about what people think or say about me."

Obviously, any man who wins three straight Golden Gloves isn't all that anonymous, but it must seem that way to Cesar in the busy Cincinnati clubhouse. Even his locker is at the opposite end of the room from the celebrities.

Bernie Stowe, the veteran major domo of the clubhouse, positions the lockers of the players so as to separate the stars — except in the case of Morgan and Rose, who were put into adjoining areas by design when Joe joined the team.

Cesar dresses at the precise opposite end where if there is a "celebrity" he is it. He likes his privacy and even his young son, Cesar Jr., does not play around the clubhouse with the same reckless abandon that the sons of Tony Perez did before their father was traded.

Geronimo is a sensitive person who accepts the nickname of "The Chief" even though he neither relishes it nor has a connection with the Indian ancestry that could link him to the lengendary outlaw Chief of the Apaches.

"I knew why I was called that," he says of the nickname. "It does not bother me. My name has no special meaning but I suppose there are people who think I am an American Indian.''

It is worth noting that Geronimo, although considered a ''throw in'' on the Morgan deal, was very high in Bob Howsam's considerations for his team in Riverfront Stadium. Bobby Tolan had been the Cincinnati centerfielder in 1970 when the team switched from smallish Crosley Field to the more spacious dimensions of Riverfront. When Tolan ruptured an Achilles tendon in 1971 and did not play a single game, the Reds went through the season groping for a man to fill the position.

''We wanted speed and defensive strength,'' Howsam said, ''and while Morgan definitely was No. 1, we regarded Geronimo a solid addition for those other things.''

''Geronimo is your ideal centerfielder for a big park, like we have all around the league now,'' Anderson says. ''How many times have you seen him outrun a ball hit up the gap on either side? He saves us a lot of runs. Maybe he isn't the best I've ever seen getting a jump on the ball instinctively, but he reacts very well.

''The other thing that makes Cesar so great out there is that he never shies away from the fence. And somehow, he has the knack of bouncing off of it.''

Geronimo insists that he once threw out his fellow countryman, Houston superstar Cesar Cedeno, trying to score from third on a fly ball, from 380 feet.

''He was very surprised,'' laughed Geronimo, his eyes twinkling, ''and so was I.''

The Reds and Geronimo also owe a debt of gratitude to someone not a member of the Cincinnati organization. In fact, Tom Lasorda, the new manager of the Los Angeles Dodgers, should be called an arch-enemy. Geronimo had a shoulder problem during the 1973 season and that winter found himself playing on a team managed by Lasorda when he returned to the Dominican Republic.

"My shoulder was still bothering me and I was getting back into bad habits with my swing," Cesar recalls. "He took me out to the ball park in the mornings for extra batting practice and pointed out the things I was doing wrong with my swing."

Cesar won't turn 30 until March of 1978, a statistic that indicates he should be patrolling the outfield with his graceful, long strides for a few years yet to come.

The young seminarian who used to listen to baseball games when he should have been sleeping in the early 1960's has come a long way. He may be one of the least known and least appreciated among the collection of Superstars who have made the Reds what they are, but he holds down a valuable position. The strength of baseball teams on defense is up the middle — and with the Reds that makes it Johnny Bench, Dave Concepcion, Morgan and Geronimo.

THE COWBOY HUMORIST

Take a bit of the Dizzy Dean, blend it with a dash of Will Rogers, visualize the sad face of a Johnny Cash — and what do you have?

You have a Pat Zachry, that's what.

Pat Zachry is a pitcher with the kind of a leathery face and lanky body who would have been cast in at least 100 western movies if he'd turned up in Hollywood in 1935. In fact, he has one of those one-of-a-kind faces that the late, great director John Ford dearly loved to throw into cameo roles.

Actually, he's seldom sad and his humor is as authentic American as John Wayne's saddle. It isn't what Pat says that breaks people up — it's the way he says it. He could have stood on the deck of the Titanic and drawled, "Ah'm not so sure we ought to lower those lifeboats just yet" and people would have howled.

As co-winner in 1976 of the National League Rookie of the Year honors (with San Diego's Butch Metzger), he made a significant contribution to the championship year the Reds put together. He climaxed his 14-7 year by pitching a victory over the Phillies in the championship series and followed that with a win in Game Three of the World Series sweep over the Yankees.

Not a bad season for a pitcher who saw several of his teammates pass him moving up the ladder in the Cincinnati minor league system. In fact, the Reds didn't keep the 6-foot-5 Texan until his options had expired and they had no choice but to give him a chance to see how he would fare against big league hitters.

He was no highly publicized "phenom" and started off by being referred to as a "long relief" man, which means pitching when the starters take early showers, an unenviable assignment even on a team as potent as the Reds. But once again Sparky Anderson found himself in need of starters and Zachry had made a strong six-inning relief appearance against Montreal. So he was given a start in Chicago, where the Reds went on a rampage and piled up 14 runs to make his first major league victory an easy one.

Next came a fateful start against the Dodgers, a team the Reds take keen delight in beating and must beat to achieve their championship objectives. Pat had a rocky first inning and breathed a sigh of relief as he reached the dugout, having yielded two runs. Manager Anderson eyed him cooly and astonished his pitcher with a tongue lashing.

"He really jumped me," Zachry said. "He told me I was a big league pitcher but I wasn't pitching like one. He said, 'You have to walk out there like you mean business and take charge' and I guess he was right."

What Sparky hadn't been aware of was that the tall pitcher had spent most of the hour before he started his warmups visiting one of the clubhouse toilet stalls to throw up. It's a nervous habit that Pat hasn't completely overcome yet.

At one stage late in the 1976 season Anderson seriously toyed with the idea of letting Zachry open the playoff series in Philadelphia as a surprise starter after announcing that Don Gullett would pitch the opener.

"It wasn't fair to anyone to play games like that," Andrson rationalized later. "And besides, Pat would have had to know he was going to pitch sometime. He just couldn't walk out there without warming up. He'd throw up somewhere."

Zack became a "Dodger killer" and ran up five straight victories over the Reds' chief Western Division rivals until they finally caught up with him in his last outing. They "mauled" him to win 2-1 behind lefty Doug Rau.

Manager Anderson takes no credit for being able to predict Zachry would emerge as a topflight pitcher.

"He has shown me that sometimes you have to look beyond the man himself," the manager said. "You look at him in the Instructional League and you say he isn't overpowering and you'd rate him a very marginal big league prospect.

"But he can throw strikes and he is competitive. So you put him in Riverfront Stadium with a Johnny Bench catching him, a great infield and outfield behind him, and Mr. Zachry is smart enough to realize what he has going for him. Now, you've got a different view of the man.

"Guys ought to pay to pitch for this team if they have the ability and any confidence in themselves," Sparky adds.

It sounds incongruous, what with his need of Rolaids and Alka Seltzer to calm a nervous stomach on the days he pitches, but Zachry thinks he "relaxed enough" under the guidance of Manager Vern Rapp at the Indianapolis farm club in 1975 to be ready to clear his final hurdle the following year.

"I wasn't emotionally ready for the big leagues in 1975," he confesses. "I didn't know it though. I was trying to be too fine, make every pitch a strike on the black (corner of the plate) at the knees. I had to relax, and just try to throw strikes and I never thought of pitching quite that simply. Oh, I know there's more to it, but basically that's what you do."

Zachry pleads guilty to being one of those clumsy type kids who was always falling over somebody or something. "I could knock a chair over getting to the dining room table.

"And when I was a kid, everything in the world happened to me. If anybody was going to get into trouble, it would be me. Like diving out of a tree and splitting my head open. Oh yes, I broke my head open. Oh yes, I broke my ankle that way too. I banged my head into the backboard in basketball practice."

Zachry's fondness for practical jokes, often at his own expense, is well remembered around Richfield High School in Waco, Texas.

"We'd be going to play a game," says Ed Knipper, who was his first high school coach, "and everybody would be serious, all intent on baseball. But Old Zach would be over there with his cap on sideways, throwing behind his back and he would just have tied somebody's jacket into knots."

Zachry also had a habit of applying "red hot," that well-known powerful analgesic salve, inside a teammate's glove or on portions of his uniform. It was a triumph for him on the occasion he could sneak some inside an unsuspecting victim's athletic supporter without the poor guy discovering it — until too late.

There also is an episode that happened to Pat when he was 15 years old that he refuses to discuss, involving the classic story of an irate father with a shotgun going after a young man he suspected of establishing a too-intimate relationship with his daughter. This father, chasing Zach, pulled the trigger and sent a blast into the young man's back, no trivial incident. It was a close brush with becoming permanently disabled.

It's hard to believe that a youngster as lanky as Zachry could be envisioned as a catcher, but the first baseball glove his father bought him was a catcher's mitt. Years later he disclosed this during a World Series press conference in New York's Americana Hotel, where he upstaged both his own manager and Yankee manager Billy Martin with his homespun wit. Asked what discouraged him from pursuing his career as a catcher, Pat responded, "Foul tips."

With the National League permitted to use a designated hitter for the first time during the Series, he was asked if he would miss the opportunity of going to bat against the Yankees and Dock Ellis. "I will," he quipped, "but I don't suppose Sparky will."

There is a poignant footnote to much of Zachry's humor in that session with more than 200 baseball writers and sports columnists from all over the country.

He talked at length about his pet German Shepherd dog, Bronco. Bronco had been with Pat all through spring training and stayed in his apartment in Cincinnati during the season. When the Reds went on the road, Bronco either was boarded in a kennel or spent time at the Pete Rose household when Karolyn Rose volunteered to keep him — which she did on several occasions. "He's just like me, neither of us can gain any weight," Zachry said of the dog.

"Sparky made me send Bronco back to Texas," Zachry drawled at the Series gathering. "He said New York was no place for a dog."

Zach's most publicized quote was his answer when questioned about what he planned to do with a winning World Series check if the Reds, ahead two games to none at that stage, did indeed win the Series.

"I'm gonna take the check to the bank and cash it,"

he said. "Then I'm going to spread all those bills on my bed, jump right in the middle of it and roll around. Later, I'll buy a couple of cans of dog food for Bronco."

What Pat didn't know, and what his mother and father who were in New York didn't tell him, was that Bronco already had disappeared from the yard of some relatives who were keeping him.

"We tried very hard to find him, offered a good reward and all of that," Zachry said after he had returned to Waco. "A lot of people tried to help us find him. But he's gone. I really was heart broken."

Upon hearing the story of Zachry's dog in New York, Cleveland Indians' General Manager Phil Seghi, who was farm director for the Reds for nine years, was reminded of a call he got once from a minor league manager about one of his pitchers.

"We really thought this kid had a future as a pitcher," Seghi recalled. "The manager said the kid was calling home too often. I asked him what was the matter with a kid calling home and he told me all he wanted to do was talk to his dog."

In pitching terms, Zachry had mastered the changeup, a trademark among pitchers developed in the Cincinnati farm system, to augment his fast ball.

"I saw Zachry pitch five years in our system and, to tell you the truth, I never thought he'd make it," says Ray Shore, one of the top Cincy scouts. "But then his last year in Indianapolis he was challenging hitters. It makes all the difference in the world."

There is a suspicion among the people in the organization at this writing that Zach's future might be enhanced if he gets a little "more mean" on the mound. Translated, that means he's going to have to brush hitters back off the plate with more regularity.

"I know a man has to be aggressive up here or they'll walk over you," Pat concedes. "I didn't use to think that way, but that's what I learned at Indianapolis my last year down there. I had to ask why Rawly Eastwick and Pat Darcy and Will McEnaney had all gone up and I was still down there. Vern (Rapp) convinced me I had to push myself to really take charge.

"That's what Sparky got all over me about that first game I pitched against the Dodgers. I still have trouble in the first inning a lot of times, but it isn't because I'm not giving it everything I got. I still am doggone nervous when I walk out there. I can't help it."

Despite his flippancy, Zachry is a very sentimental person. "When we won the pennant and my father came in the clubhouse, I put my arms around him and started to cry like a baby," he confessed. "I couldn't hold it back."

He wasn't crying earlier in the year when the Reds were playing an exhibition game at Indianapolis. The starting lineups were being introduced, with Manager Anderson out on the field first followed by several players, when Zach yelled at an unsuspecting Jack Billingham, "Jack, they've just called your name. Get out there."

Billingham ran out, found himself standing next to Johnny Bench who had just been announced as the cleanup hitter. "What the hell are out doing out here?" inquired the surprised Bench.

Going into the 1977 season, there were wedding bells in his future plans — a school teacher from Houston, Texas having come into his life. If indeed he does get married, he may have to give up one of his favorite lines, "There ain't no place like home, after all the other places close."

He has paraphrased that to "there ain't no place like the big leagues especially if you've seen Sioux Falls and Three Rivers."

Doug Flynn greets infrequent slugger Geronimo

Cesar takes "low road" under Enos Cabell, so does coach George Scherger

That's Zachry's unclassic "X" follow through

The smooth centerfielder

THEY ALSO SERVED

Ty Cline hardly enjoyed what could be called a meteoric major league career, but when he retired it was on a high note.

Tyrone Alexander Cline, of the South Carolina Clines from the historic city of Charleston, was one of the "spear carriers" in the championship production of the Big Red Machine in 1970. He took a very fortunate bounce from the lowly Montreal Expos to the Reds on June 15, the annual trading deadline, to fill the need for a lefthanded pinch hitter and a competent backup defensive outfielder.

When the Reds took on the Pittsburgh Pirates in their first post-season championship series, Gary Nolan and Dock Ellis matched excellent pitching strides through the regulation nine innings. The game was scoreless moving into the 10th. Leading off that inning, Cline was sent up to bat for Nolan and responded with a triple up the left center alley in Three Rivers Stadium. Before the inning was over, the Reds had scored three times with Ty's run the official winning tally as such things are computed.

In the third and final game, the winning run in Cincinnati's 3-2 contest was carried across the plate by Cline, this time as a pinch runner. The Reds scored nine runs in the three games and handyman Cline tallied the first and last. In the somewhat disastrous World Series against the Baltimore Orioles, Cline delivered a single in three pinch-hitting appearances. When the Reds were cutting their squad in spring training the next year, Cline's age was a factor in putting him on waivers. Manager Sparky

Anderson called him into the office to give him the bad news.

"That's all right," responded the 31-year-old outfielder. "It's time for me to hang 'em up. I've got a business to take care of."

And so the whippet-thin outfielder who put in time with Cleveland, Milwaukee, Chicago Cubs, Atlanta, San Francisco and Montreal before getting to Cincinnati, went back to Charleston to dip chocolate almond and banana royale ice cream. He had used his World Series check for a down payment on a Baskin-Robbins franchise.

Cline is one of several dozen athletes who passed through the revolving door in Riverfront through the 1970's. He was one of the dozen members of that 1970 World Series eligibility list no longer around by the time the Reds faced the Oakland A's in the Series 24 months later.

Cline is worth remembering not only because of his contributions to the cause, but because he had the temperament to fill a utility role to near perfection. By the time he reached the Reds, any thought of being a day to day player had been extinguished by the school of hard knocks. He had made the necessary mental adjustment to accept his role and be ready to execute it when called upon.

It would be impractical to list all the "extras" who are known by various other nicknames in the clubhouses — most of the names being unprintable. One that is printable is "scrubini," a term with its origins in "scrub team." But the Reds have thrived on them with the likes of Doug Flynn, Mike Lum, Bob Bailey, Ed Armbrister and Bill Plummer being prominent in the picture in the 1976 championship season. Also in that lineup was Dan Driessen, then still a backup man to Tony Perez.

Over the years, Darrel Chaney, Woody Woodward and Jimmy Stewart were invaluable as infielder backups. That trio had much more longevity than most of the extra outfielders, although the role of being Johnny Bench's backup catcher has been an interesting one for Pat Corrales and then Bill Plummer.

If you're interested in posing a trivia question, Bench was not behind the plate for the first four games of his rookie season in 1968. Don Pavletich, who had come into the Cincinnati organization as a bonus baby 11 years before and had six full seasons as the backup to John Edwards, was in the lineup. He had enjoyed such a strong spring showing that Manager Dave Bristol felt he deserved the opportunity even though Bristol was head over heels in love with Bench — as an athlete.

California-born but of first generation Mexican parentage, the thick-torsoed Corrales had been acquired by the Reds prior to the 1968 season but farmed out to Indianapolis. He had played briefly with both the Phillies and Cardinals before he moved into what was to become a three-plus season period as Bench's understudy.

"I hated the idea of accepting the fact I would never be more than a No. 2 man," Corrales said after adjusting to his role. "But one thing was for certain. I knew I could never go to the manager and say 'what has Bench got that I haven't got?' and so in that sense it was a break for me.

"From that time on, I started gearing myself to try to make a fulltime career in baseball as a manager or a coach," explained Pat, who is a coach with the Texas Rangers at this writing in 1977.

That same adjustment came later for the husky Plummer, like Corrales a Californian, who found himself working behind a man many considered as fine a catcher as ever lived.

"It isn't easy to tell yourself that you're completely stymied," Plummer said several years later, expressing much of the theme Corrales had talked about. "I'd like to think that someone, somewhere would have confidence that I could be a regular catcher. But I've been waiting for the National League to expand and that hasn't happened."

Corrales and Plummer are convincing examples that physical strength alone does not make one a good hitter. In terms of strength, both men wouldn't take much of a back seat to Bench — but Corrales, who had trouble hitting breaking pitches, had only four big league home runs in nine seasons.

An agile, 200-pounder, he would be a guy you'd want on your side in a gang fight but his batting swing against breaking pitches was all too delicate.

In the mid-1950's the Reds had an outfielder named Pete Whisenant who could have been a .400 hitter if he had faced nothing but lefthanded pitchers throwing fast balls. But Pete could not cope with a righthanded curve ball, despite a set of massive arm muscles. One day in a rage, he flexed his right arm into a position that would have done Charles Atlas proud and profanely snorted, "Bleep, bleep, all that bleeping strength and you can't hit a curve ball."

Plummer, unlike Corrales, has savored at least one unforgettable day at the plate. On June 5, 1976, he drove in seven runs in a 12-5 romp over the St. Louis Cardinals with a single, bases-loaded triple and a three-run homer.

One Cincinnati "scrubini" who talked his way off the club and struck it rich was Hal McRae. As a member of the Kansas City Royals, he missed out on the American League batting championship in his last official at bat of the 1976 season.

McRae had never adjusted to the bench-warming

philosophy mainly because he felt cheated by fate. He had been a real speedster seemingly ticketed for a quick rise to the majors after hitting .295 at Indianapolis in 1968. But a couple of months later in winter ball in Puerto Rico he nearly ended his career, suffering one of the worst multiple-fractures in his right leg that doctors ever attempted to restore to competitive ability.

Hal hobbled through 1969 on a rebuilt leg into which a 10-inch aluminum spike had been inserted as a support for the largest bone. The spike was removed prior to the 1970 season, but McRae's mobility was still seriously limited as he divided time in left field with Bernie Carbo.

Demoted to utility duty in 1972, McRae's cries of indignation helped motivate a winter trade that sent him to Kansas City where, within a year, the American League's adoption of the designated hitter rule gave him a new lease on life. Moreover, with the passing of time the leg has gotten stronger and Hal can play his position defensively without being the kind of risk Sparky Anderson felt he was when he appeared in 99 games in 1971.

Most athletes with the ability to reach the major leagues have been high school, sandlot or college stars, and a great many of them were multi-sport celebrities in their schools and home towns. Hence the ultimate acceptance of a humble role is tough on the ego or, if you will, the normal "macho" in 99 per cent of the masculine performers.

Darrel Chaney became a "liberated" scrubini when he departed the Cincinnati scene to become the regular shortstop of the Atlanta Braves in 1976 — at age 28.

The likeable Hoosier was a high school quarterback of considerable renown in Oliver Morton High School in Hammond, Indiana, and a guard with a 15-point average in basketball. Football coaches of the caliber of Notre

Dame's Ara Parseghian and Michigan State's Duffy Daugherty came to Chaney's home in an attempt to recruit him. Darrel opted for baseball instead, being the No. 2 draft pick of the Reds behind Gary Nolan in June of 1966.

Chaney came up to the Reds in 1969, having played the previous season on the Asheville, North Carolina team managed by Anderson. The Reds had traded veteran shortstop Leo Cardenas to the Minnesota Twins for pitcher Jim Merritt and the battle for shortstop was between Chaney and Woody Woodward, who had come over from the Braves in a trade the previous June.

Darrel found himself overmatched with the bat against big league pitching, but played his position well enough to get into 90 games. His first year batting average was .191, one pound over his playing weight.

"On opening day my Mom and Dad drove down to see me play my first major league game," he recalled. "I was 21 years old, a big leaguer, but I never got out of the dugout. I really felt strange."

A year later, David Concepcion was on the scene. The shortstop battle started off as a three-man affair but Chaney came out on the short end. He appeared in only 57 games, went to bat 95 times. This was what would normally have been his senior year had he taken a football scholarship. Would they have loved him in South Bend or East Lansing?

But, as the vaudeville joke goes, things could get worse — and they did. Banished to the Indianapolis farm club a year later, Darrel swallowed his pride, had a good year and bounced back.

"The toughest thing about going back to Indianapolis was for me to concede that I had to start thinking about my baseball future in a different light," Chaney said. "Here I was, a 6-foot-2, 190 pounder who was never going to be a power hitter. I had to start thinking like a little man and the guys called me a 'Judy' hitter and I had to learn to ride with that because that's what I had to be."

It wasn't that Chaney had resigned himself to becoming a utility player, merely the acceptance of doing what he could do within the limitations of his physical talents. Physically, he was slightly taller and just as heavy as Hank Aaron — but the latter was crossing the 700 plateau in home runs before Chaney hit No. 7.

Chaney went to spring training as late as 1973 with the feeling he had a chance to beat out Concepcion for the starting shortstop job. This was hardly out of the question in that the Venezuelan's batting average had slumped to .209 the year before.

"But Davey came on like gang-busters," Darrel said later, "and by the time he broke his ankle in July and I had the opportunity to play, I knew in my heart he was a better player than I was."

Chaney had the candor and decency to say that publicly, and by that time Darrel had become an extremely popular Red. His mates knew him as "Norton" — the nickname being based on the character Ed Norton as played by Art Carney on television.

"I'm not resigned to being a utility player my whole career," he said in 1975. "With this club, I am. What can you see when you're trying to play behind Joe Morgan, Concepcion and Pete Rose?"

By the time he was traded to the Braves, Darrel had grown some deep roots in Cincinnati. "I got into three World Series with the Reds and we finally won one," he said. "I'll have an opportunity to play in Atlanta and I'll appreciate that."

In his first two years with the Reds, Chaney roomed with Woodward, whose career developed along similar lines. A graduate of Florida State University, where he eventually returned to settle down happily as baseball coach, Woody had been a bonus baby with the Braves when they were still in Milwaukee. He spent the 1964 season as a seldom-used benchwarmer. When he had the opportunity to play regularly, both with the Braves after the team shifted to Atlanta and in Cincinnati, a weak bat prevented him from ever nailing down a permanent job.

"I was able to help Darrel in his early years because I knew his frustrations from first hand experience," said Woodward, who had been an All-American collegiate shortstop.

Woody was only 29 years old when the opportunity to return to FSU came his way. "I knew it was time to make my move," he said. "I had opportunities to make it in the majors and I never lived up to my own expectations. I didn't want to hang on and bounce around with this club and that."

Strangely, Woodward enjoyed an unusual degree of success hitting against Tom Seaver, one of the finest pitchers in baseball and the mainstay of the New York Mets.

Chaney became vulnerable in Cincinnati only because Doug Flynn played his way onto the club in 1975 with all-out hustle and a "yes, sir" attitude that delighted Anderson.

Doug's arrival on the scene with a team destined to win a World Series was a long-shot, a term young Flynn would be familiar with having grown up in Lexington, Kentucky, that hotbed of thoroughbred racing and breeding. He wasn't drafted by any team the year he graduated from high school and had few college scholarship offers. He was a "walk on" on the University of Kentucky freshman basketball team but wound up playing at Somerset Junior College. He was signed by the Reds at a tryout camp, and started with minor league batting averages of .211, .258 and .253.

"I had never thought about a career in baseball," he said "When you grow up in Lexington, you want to play Kentucky basketball. Heck, I didn't even expect to be invited to the Reds training camp in 1974. I was working as a stock boy in a supermarket when the letter came."

Flynn's father, Bobby, who had played basketball at the University of Dayton, had become a top flight college basektball official, an avocation he could pursue while running a successful insurance business.

"We were worried about Doug's hitting," conceded Anderson. "But I keep forgetting I've got the greatest batting instructor in the world in Teddie (Kluszewski) and when he gets a willing kid like Flynn, he can really help him."

Flynn also had the good fortune to be "adopted" by Pete Rose — who would have felt so much more welcome as a rookie himself had he been similarly treated.

"He's super, just plain super," Flynn said at the time. "He really went out of his way to make me feel I belonged. A rookie can't help but be a little awed his first few weeks in the league and Pete always seemed to be around to tell me the little things I didn't know about doing, like how to tip around the hotels."

Few rookies have hit the jackpot as Flynn did, with two World Series shares in his first two seasons, plus some other off-field revenue. Even so, he faced a tough decision with the free agent option in the 1976-77 off-season.

"I was tempted not to sign, being behind Joe Morgan and Dave Concepcion as I am," Flynn put it candidly. "I'd like to go somewhere and see what I could do playing every day. If a fellow has to be a utility man though, I couldn't think of a better team to be with. But it's only natural to have confidence in your ability to play regularly."

Bachelor Flynn's value in a secondary role is best stated by Rose. "When Doug is at second, he looks like a second baseman, when he's at third he looks like a third baseman, and when he's at short he looks like a shortstop," says the veteran who has played so many different positions himself.

Easy-going, guitar-strumming Flynn has shown an affinity for communicating with youngsters and has positive evidence in helping turn the lives of at least two troubled persons. Flynn, George Foster and Dan Driessen visited a "halfway house" in Dayton where they were participating in an annual clinic sponsored by the Dayton Journal Herald during the All-Star game break.

"This one girl jumped all over me," Flynn says, telling the story. "She wasn't going to let some big shot like me tell her what she should be doing.

"I don't really remember what I said to calm her down, but I stayed around after talking to the group and talked with her and her attitude was changing. I promised to write to her and pretty soon we were exchanging letters. She's living in a foster home now and, I'm happy to say, getting straightened out. I don't claim all of the credit for that, but it's nice to feel that I helped in some way. At least, the girl thinks I did."

The road that Bob Bailey travelled to become a Cincinnati "scrubini" could be called by far and away the most glamorous and well-compensated one. As a California high school phenom, Bailey banked a cool $165,000 from the Pittsburgh Pirates the day after he graduated when he could be signed on the open market, this being ahead of the free agent draft.

Bailey played his first big league game in September of 1962 — earlier than any other person who was with the Reds when he joined the team going into the 1976 season. In his four seasons with the Pirates, two in Los Angeles and seven with the expansion Montreal Expos, Bailey was a proven big league hitter with good power — especially against lefthanded pitchers. Defensively, he wasn't much above average as a third baseman and his relatively slow speed makes one wonder in retrospect what happened to the stop watch the Pirate scouts were using to time him.

Bailey has to be the highest priced "scrubini" the Reds have ever had, and deservedly so on his ability as a pinch hitter and a valuable fill-in.

"There would be a world of difference riding the bench in Montreal and in Cincinnati," Bailey said of his joining the Reds. "With this team, you know why you're here and you adjust. In Montreal, I would have felt I should still be in the lineup. The trade to Cincinnati made me conscious of the fact I had a shot at the World Series and it turned out that came true."

Although not nearly as highly paid during his career, Mike Lum looked upon his being traded to the Reds in the same light. Unlike Bailey, the 31-year-old Hawaiian-born Lum was regarded as excellent defensive insurance for both outfield and first base play.

One thing Lum found about not being in the lineup every day was that keeping his weight down became a problem. To come in under his listed 180-pound weight, Lum spent much of the winter of 1976-77 learning to scuba dive off Waikiki Beach.

Among the more interesting outfielders who played brief stands in Riverfront were Andy Kosco, Gene

Locklear and Richie Scheinblum, a diversified lot.

Bespectacled Andy Kosco is a player who deserved a kinder smile from fate than he got. An always smiling, good-natured and sincerely compassionate individual, Kosco made the rounds of the majors and minors always giving his best and hoping for the big break-through that never came.

Every manager who ever had him put a young player needing some friendly guidance on the list as his roomie. When with the Dodgers, he helped break in Bill Russell, a country boy from Kansas. The father of five youngsters, Andy had plenty of practice as both a father and big brother.

Kosco's total contribution to the Reds was 47 games in 1973 and 33 the next year before a chronic bad back influenced his decision to give up and finish his work toward a law degree.

Locklear is a full-blooded Lumbee Indian out of North Carolina who, naturally, was called "Chief" in his baseball travels, a fate that Gene bore with the stoic patience of his ancestors. It probably never occurred to the people calling him that that they were dealing in a racist generality.

Locklear and Schienblum had something in common. Both wielded a bat from the left side of the plate well enough to warrant a job in the majors. Alas, both were strictly a liability afield although neither ever accepted that judgment despite the unanimity among the managers and scouts who assessed it.

"Locklear stalks a fly ball like he's going after a wagon train," quipped one observer.

Gene also has a God-given talent as an artist. His paintings and water colors have received critical approval when exhibited and Pete Rose owns several pieces he purchased from the congenial Locklear.

Gene's prowess as a hitter is evidenced by the fact he led the Eastern League and American Association in hitting in 1971 and 1972 respectively. Traded to San Diego for pitcher Clay Kirby, he hit well for the Padres but his indifferent play afield kept him in manager John McNamara's doghouse and he was waived late in the 1976 season to the Yankees who wanted him as a pinch-hitter.

Schienblum had bounced around with four American League clubs before making a brief stop in Cincinnati during the 1974 season. New York-born Richie always was good for a story and kept a wary eye out for the baseball writers, a habit that got him a lot more "ink" than his ability warranted. This was especially true in his first stop in Cleveland where he was a perennial springtime hopeful.

Panamanian Ed Armbrister stands a chance of being the guy with the lowest ratio of playing time per World Series check in baseball history. Ed was "No. 5" among the players acquired in the famous deal with Houston in which Joe Morgan was "No. 1." For all of that, Sparky Anderson talks about him as an "untouchable" as a speedy pinch runner, defensive replacement or bunting specialist.

It was Armbrister's alleged interference on a sacrifice play that ignited the fireworks in the 10th inning of the third game of the 1975 World Series. Armbrister, laying down a bunt batting for Rawly Eastwick, collided with Boston catcher Carlton Fisk. Fisk grabbed the ball and threw it into centerfield trying to nail Cesar Geronimo at second.

Had the throw been accurate, the Red Sox might have had a double play because Armbrister was late getting going for first base. The wild throw, however, put the winning run at third base. The Red Sox howled but American League umpire Larry Barnett ruled it was not interference. Joe Morgan scored Geronimo with a base hit a few minutes later.

The argument over the play lasted most of the winter and still hasn't subsided entirely in New England. There was no such argument in the 1976 Series sweep over the Yankees since the dominance of the Reds was so great that Armbrister remained a spectator on call in the dugout.

No recall of Cincinnati's "scrubinis" would be complete without a mention of the late Chico Ruiz. Hiraldo Sablon Ruiz was a Cuban, one of the "class of 1959" to emigrate from Fidel Castro's island to play in the Cincinnati farm system. If Tony Perez was the ultimate in even-tempered Latins, Chico was the ultimate in laughably funny even-tempered Latins. He put in six full seasons in Cincinnati uniform (1964-69) before being traded to the California Angels. He was killed in an automobile accident February 9, 1972.

Ruiz somehow managed to create a character role for himself that he was perfectly content to sit around and watch everyone else play. This was hardly the case, at least in the beginning, but legends are unpredictable. He became known for carrying around his seat cushion, suitably autographed by the likes of Willie Mays, Bob Gibson, Roberto Clemente and the other superstars of that era.

On the field, Ruiz contributed one incredible bit of daring that nearly paid off in the Reds "stealing" the 1964 pennant. That was the year the Phillies had the pennant seemingly sewed up, only to have the Reds charge past them and finally it was the Cardinals who won.

In the first of a three-game series in Philadelphia, with

the Phillies 6½ games in front of the Reds with only 13 to go, Ruiz found himself in the lineup. Perched on third base with one out in a scoreless game in the sixth inning, he watched pitcher Art Mahaffey concentrate heavily on dangerous Frank Robinson at the plate. To the astonishment of everyone in the park Ruiz broke for the plate on a Mahaffey windup and stole home. Slugger Robinson could have fractured his skull with a swing of the bat, but Chico was safe and that was the only run of the game. That 1-0 Cincinnati win started the Phillies on one of baseball's more incredible tailspins, a 10-game swoon that cost them the pennant.

That was before Chico made himself a character, leading up to his classic line to Manager Dave Bristol after being in the lineup several days in a row.

"Bench me or trade me," Ruiz told Bristol, paraphrasing the demand of hundreds of players over the years to be put into the lineup or traded.

The most dramatic feat by a Cincinnati "scrubini" of the current era, and perhaps ever, belongs to Hal King, a moon-faced catcher who went to the plate exactly 63 times in parts of two seasons. King held such a hot hand in mid-season of 1973 that he should have gone to Las Vegas instead of reporting to the Reds when he was called up from Indianapolis. At that time, using his luck at the dice tables, he might have wound up owning Howard Hughes.

When King was coming up in the Houston organization in the mid-1960's a prominent member of the Astros was quoted as saying that he wouldn't trade him for Cincinnati's Johnny Bench. The man who made that miscalculation is a nice enough person to be worthy of protective anonymity.

King's defensive abilities were minimal at best and his motivation questionable. After the Astros dropped him, he was briefly with Atlanta and Texas before being picked up by the Reds at the minor league level.

The Reds were seemingly going nowere fast as of June of 1973 faded into July and the Dodgers had built a whopping lead. In fact, by winning the first of a four-game series in Riverfront on June 30 they moved 11 games ahead of Sparky's men in the Division race.

There was a doubleheader the next day and in the first game classy Don Sutton held a 3-1 lead going into the ninth inning. With two Reds on and two out, King was sent up to bat for pitcher Fred Norman.

King's only previous at bat in the six days he had spent with the Reds had seen him deliver a pinch home run. Lightning was destined to strike again. Sutton made a mistake, getting a fast ball in what would be called Hal's wheelhouse, meaning the upper part of the strike zone where he had a full cut. He hit a towering, deep smash toward the right field corner. The only question about it clearing the playing area was whether it would be fair or foul.

The ball struck the screen just inside the foul pole and it took a moment for most of the crowd of 46,047 to realize they had just witnessed a dramatic game-winning home run. No one knew it at that moment, but King's blast had turned the National League season around. The Reds easily won the second game, took the series finale from the Dodgers Monday night, and wiped out the Los Angeles edge by early September to win going away.

But King wasn't through rolling sevens or filling inside straights. He hit a pinch grand slammer against the Expos eight days later and beat the Mets with his fourth (and last) pinch hit home run August 17.

It was, however, a brief fling of glory. When Hal had only three hits in 17 at bats midway into the next season, he was sent back to Indianapolis and hasn't been heard from since at the major league level.

Because the Mets ambushed the Reds in the playoff for the pennant, Hal's heroics didn't even net him a World Series ring. And if the Reds, in a moment of compassion, ever invite him to an Old Timer's game, most of the spectators will have to be reminded of why he's there.

Victor, Pituka, Tony and Eduardo Perez become U.S. citizens

Mets' Duffy Dyer misses ball, Tony misses plate

THE BULLPEN

"Did you ever hit anybody in the head with a ripe tomato?" asked Clay Carroll of no one in particular.

"It sure makes a funny noise and there's a way it flattens out to really mess up a guy's hair," he went on, undaunted by the fact his listeners didn't pick up the conversation.

This gem was reported by an eavesdropper sitting behind the Reds' bullpen crew in Pittsburgh one day merely to record for posterity what goes on in that isolated baseball outpost.

The simple pleasure of drilling a buddy with a tomato is a memory from Clay's days of picking cotten in southern Alabama. Since cotton pickers in the deep south are either black or very poor, this fact provides a clue to Clay's boyhood.

"The Hawk," as Carroll liked to describe himself, was perched in the Cincinnati bullpen from June 1968 until he was sent to the Chicago White Sox in a mysterious trade at the end of the 1975 season — the mystery being that the Reds had decided to unload the aging relief pitcher and his fat contract and were not fussy about what was offered in return.

This ignominious departure was hardly a just reward for Carroll's years of service, and the big Alabaman is a fit subject to launch a study of the bullpen which has been a prominent factor in the Sparky Anderson years in Cincinnati. Strong relief pitching is almost a pure necessity to win pennants in the modern baseball era where the rigors of travel, night games and the lively ball itself make complete games relatively rare.

The origins of the term "bullpen" are hazy and theories are numerous. As good as any is the one claiming that it was so named because the Bull Durham Tobacco Company frequently had advertising signs along outfield fences, the usual location of bullpens.

In the oddly-shaped Polo Grounds, for many years the home of the New York Giants, the bullpens actually were in fair territory but a good 400 feet from home plate in deep left and right fields.

There are no bullpens in the strict sense in Cincinnati's Riverfront Stadium, as well as in some other new parks. The "bullpen crew" is in the dugout itself and runs a few steps out to the warmup area when it is needed.

The fanciful stories from the bullpens are from the parks and times when they were isolated outposts, with communication mostly by telephone.

"The first thing the guys usually would do when we walk out there to make ourselves at home," says Cincinnati catcher Bill Plummer, "is settle down for a serious look around the stands to find the attractive girls. Even if a guy is serious about the game, you're pretty far away and the early innings generally aren't all that interesting."

In many ways Clay Carroll can be called a typical relief pitcher. The man often as not comes into a game in a critical situation in which he has no margin for error. In that sense he has to be completely self confident of his ability to challenge hitters.

"I can't mess around with being cute," Carroll put it himself. "I come in humping and pumping."

"The Hawk" was a fascinating blend of physical strength, native shrewdness, honesty and an almost childlike fascination for being appreciated and recognized. In fact, it was Dave Bristol's astute grasp of Carroll's personality that converted him into a standout pitcher. Bristol pumped up the Carroll ego in so obvious a manner that a lot of athletes might have been turned off by such an approach.

"You can really bring it, Hawk," Bristol would say to his man when he was summoned in a tough situation.

The Atlanta Braves, who had signed Carroll in 1961 when they were still the Milwaukee Braves, apparently never had anyone who understood what it took to motivate the big raw-boned southerner. In his first two months with the Reds under Bristol, he notched five saves, four victories and had one stretch of 17 scoreless innings. He came into his own nationally with his fine work in the losing cause of the 1970 World Series, being credited with Cincinnati's only victory over the Baltimore Orioles.

When Carroll came to the Reds, the ace of the bullpen was another Southerner, lanky Ted Abernathy, who drove a fuel oil truck route in the small North Carolina town where he lived in the off-season.

"Abby" was a side-arming sinker ball pitcher who survived some of the most precarious situations with an unflappable coolness that defied explanation. There was the night in Atlanta Stadium when he came in with the bases loaded, one out in the ninth inning and the Reds clinging to a one-run lead. All he had to face was home run king, Hank Aaron, followed by Mack Jones, a dangerous lefthanded hitter.

When Abby ran a 3-0 count on Aaron, another ball would have forced the tying run across. But he threw a called strike, got Aaron to pop up on the next pitch, and then fanned Jones.

Asked what he was thinking about while pitching 3-0 to Aaron, he replied, "I looked into the dugout and saw

Dave's face. That was all the message I needed."

It was the development of Carroll as a workhorse and Bob Howsam's acquisition of slender Wayne Granger from St. Louis that made Abernathy a surplus commodity and hastened his exit to the Cubs. One of the players acquired in that deal was catcher Plummer, destined to spend many an hour in the Cincy bullpen over the years.

Granger, a very slightly built athlete whose apt nickname is "Stick," threw a sinking fast ball with a three-quarter motion which he could drop to sidearm. But his motion was never as unorthodox as Abernathy's all-out delivery from the side.

When the Reds moved into Riverfront and played on the fast-running artificial surface, sinker ball pitchers found themselves less effective and Granger soon departed, leaving the hard-throwing "Hawk" king of the bullpen until the arrival of Rawly Eastwick in 1975.

Carroll may be the only baseball player ever to lose a decision to a swinging door — and it was not one to a bar. It happened in Los Angeles when Bill Buckner, then of the Dodgers, slid into Joe Morgan at second base. Because Buckner's spikes were high enough to cut him in the chest, Joe took a swing at the outfielder.

As always, a mob scene quickly developed. Carroll, in his excitement to get to the scene of the action, started to dash through the bullpen door in the left field corner, only to discover it securely fastened. The impact of the 205-pound pitcher tore the door off one of its hinges, but not the other. It rebounded sharply, cutting him across the cheek and stunning him.

"I thought I had been blind-sided," Carroll said later of the black eye and bruised cheek he acquired in the freak mishap.

Several years later, Pedro Borbon tore down the

bullpen door in Shea Stadium when the Reds and Mets responded to the scuffle between Pete Rose and Bud Harrelson.

If there is an unsung hero in the cast of the Big Red Machine in Sparky's era, it could well be Borbon. There have been strong-armed pitchers over the years who could throw the ball with greater velocity than this native of the Dominican Republic, but none with more resiliency in the arm itself. Because of his incredible physical strength, there has never been the slightest doubt in recent years about Pedro's ability to hold his place on the roster.

"I honestly believe he could pitch every day for 14 or 15 days at a run if you have to have him," says Manager Anderson. Starting with 1972, Borbon has averaged 125 innings over the five full seasons he has been a part of the club. While consistency has not been his greatest asset, there never has been a more "available" pitcher in the history of the game.

Pedro is not a man to tamper with. During a fight between the Reds and Pirates in 1975, pitcher Daryl Patterson was trying to pull people apart and made the mistake of grabbing Pedro — who promptly bit him.

That bite led to the Pittsburgh team physician directing that a precautionary tetanus shot be administered. Bob Prince, the Pirate broadcaster, started calling Borbon "Dracula" and someone called Pedro's attention to it. The next time the Pirates were in Cincinnati, Borbon approached Nellie King, Prince's partner on the air, under the impression he was confronting the man who had verbally maligned him.

King thought he was off the hook by identifying himself but Borbon insisted he deliver a warning for him to Prince. When King flippantly declined, saying he was no messenger boy, Pedro suggested that the broadcaster was making him angry. Prudence taking the course over valor, onetime pitcher King agreed to be Borbon's messenger.

Back in the Dominican Republic, Pedro raises gamecocks, those sharp-beaked roosters who engage in cockfights — to the death. It is a legal sport there and obviously it is not a pursuit for the timid.

Pedro's fights on and off the baseball field in the winter leagues have earned him a year's suspension and a $500 fine, among other things. Tom Lasorda, who replaced Walter Alston as manager of the Los Angeles Dodgers, was his manager one year in winter ball and went to the mound to remove Borbon. Instead of handing him the baseball as is customary, Pedro wheeled and fired the ball over the fence in dead center field.

He has collected many a $20 and $50 wager in this country that he can stand on the pitcher's mound and duplicate that feat of firing the ball over the outfield wall.

One day in the 1975 season when the Reds had one of those noon starts called the "Businessman's Special," Borbon decided to drive over to Indianapolis to see his Latin buddies on the farm team there that evening. Starting back for Cincinnati, he took a wrong turn up the Interstate highway and was halted for speeding by a state trooper just 30 miles south of the Chicago city limits, turning a 250-mile trip into one of 450 plus.

One of Pedro's strange achievements, used these days to entertain his friends in the bullpen, is an ability to flex the massive muscles in his right forearm and produce a rhythmic beat by rubbing the knuckles of his left hand along the arm.

"It isn't very tuneful," says teammate Eastwick, "but it's amazing that he can get musical sounds out of his own arm."

When one thinks of Eastwick, he also thinks of Rawly's pal, Will McEnaney. The southpaw, who grew up in Springfield, Ohio, was on the mound when the Reds wrapped up both the 1975 and 1976 World Series in Boston and New York respectively.

Known in some quarters as "Wacky Will," McEnaney was an unorthodox character. He may be the only big leaguer today whose high school coach fired him from the school baseball team as a hopeless goof-off.

Perhaps McEnaney's most unique feat in the two-plus seasons he was around the Reds was "stealing" the baseball bag from a grumpy park policeman in San Diego. This character sat on the foul line in right field guarding the ball bag, perhaps 60 or 70 feet from the Cincy bullpen in the rightfield corner. No one is quite certain why the baseballs, consigned for bullpen duty, needed guarding.

McEnaney's mischievous brain figured out that if he crawled through the corrugated pipe on which the tarpaulin was rolled, he could maneuver within a few feet of the policeman's back and the ball bag. So Will carefully crawled on all fours through that pipe, a feat of dexterity that the U.S. Army expects of its volunteers for Ranger training. With the policeman watching the game, McEnaney "captured" the ball bag and crawled back through the pipe dragging it with him. After freeing himself and brushing himself off, but with the bag still concealed in the pipe, Will casually asked the policeman for a ball.

"That cop nearly went bananas," laughed Eastwick. "He knew nobody had walked in his direction from the bullpen and it was another 100 feet into the dugout."

In Crosley Field the bullpens were along the foul lines in the outfield corners. In later years there was a small, covered dugout built in which the players could sit. The ground crew installed and kept a hose under the bench.

Marshall Bridges, a well-travelled southpaw, spent some time in 1960 with the Reds. It was his habit, if he could persuade his mates to accommodate him, to stretch out on the bullpen bench where, if not disturbed, he could easily fall asleep. One afternoon the bench was his without argument as his friends were waiting for him to start snoring.

As soon as the snoring started the hose was turned on and the small dugout was slowly converted into a shallow swimming pool. With the water lapping at his ears, Bridges suddenly had the sensation of being a passenger on the Titanic. He tumbled off the bench, gulping water, and scrambled out of his pool sputtering and yelling while hardly anyone in the park knew what was happening.

"His cap actually was floating," laughed Joe Nuxhall, a witness to the episode.

A year later Nuxhall was whiling away time in the bullpen for the Kansas City A's before Charles O. Finley moved that franchise to Oakland. The A's were playing the Washington Senators and it was a routine game in the early innings. Joe looked across the way and saw his old friend, John Klippstein, sitting very relaxed in the Washington bullpen.

Nuxhall picked up the telephone, dialed the Washington bullpen and said quickly, "Get Klippstein ready in a hurry," and hung up.

Klippstein jumped up and began heating up as if a bases-loaded emergency had developed. In the dugout Mickey Vernon, the Washington manager, didn't notice him for a few minutes but when he finally did, he blinked in amazement. He called down to ask what the hell was going on, only to be told he had ordered the frantic activity himself. The Kansas City bullpen crew found the whole business a pleasant diversion.

Such telephone chicanery is generally outlawed in a code of "unsportsmanlike conduct" and often the only numbers posted in a dugout are to the club's own clubhouse and bullpen.

During the 1972 World Series between the Reds and the Oakland A's, Commissioner Bowie Kuhn had a telephone installed at his side in one of the choice Riverfront boxes. It was a direct outside line not connected with the Reds switchboard. Somehow the number became known to a group of writers in the main press box. Legend has it that the guilty party who made the discovery was Jim Enright, the now retired Chicago writer and well known Big Ten basketball official.

In any event, the Commissioner answered a call and explained that he couldn't accept a pizza order. A few minutes later, his phone rang again and a voice wanted to know if this was the Colonel Sanders Fried Chicken outlet. By the time the Commissioner was explaining that this wasn't the Dainty Doll Diaper Service, he began to get suspicious. Knowing that the press box was the only spot in the park with access to a phone from which he could be observed taking those calls, an emissary was dispatched to plead that the line be kept open.

Over the years, telephones to the Cincinnati bullpens have been kept humming by Sparky Anderson, not at all hesitant about his growing reputation as Captain Hook. But the man has known what he was doing, as evidenced by his record.

"Complete games don't mean a thing," he insists. "I've got a great bullpen and I'm going to use it."

Use it he has, and it has been an integral part of baseball's most productive and perhaps most intersting team of its time. The story is a long way from being ended as the Men of the Machine continue to do their thing.

AFTERWORD

Just as change in one's life is as inevitable as the sunrise, personnel turnover in team sports is an accepted fact of life — even at the championship level.

"The worst crime a man in my position can be guilty of is letting his team grow old together," insists Reds' boss Bob Howsam. But no one was prepared for the flock of major changes as Howsam made several bold moves to correct a pitching imbalance just as this book was going to press.

As the June 15 trading deadline approached, Howsam was working on a blockbuster deal aimed at landing Tom Seaver from the New York Mets. Normally Seaver would have been an absolute untouchable, but he had become involved in a personality clash with Mets' board chairman M. Donald Grant.

A similar personality clash was building in Cincinnati where the unsigned Rawly Eastwick had become bitterly critical of Reds management. So Howsam worked out a deal to send Eastwick, Doug Flynn and Steve Henderson, the top hitting prospect in the productive Cincinnati farm system, to New York for Seaver.

Eastwick, however, nearly killed the deal by refusing to even enter into contract negotiations with the Mets. Frantically working against a deadline only hours away, Howsam substituted Pat Zachry for Eastwick and added Dan Norman, another outfield prospect, and the deal was made.

In spring training the Reds appeared to have an orderly

development program scheduled to bolster their pitching staff over the next several seasons. Sparky Anderson, bubbling with enthusiasm, said "we've got five of the finest looking young arms I've ever seen on the way up."

He was referring to Paul Moskau, Mike LaCross, Paul Ferreyra, Mario Soto and Dan Dumoulin. None had been figured in 1977 plans and Soto was regarded as being three years away. Moskau, college-educated, intelligent and possessor of a fine physique, nearly made the Reds change their minds in spring training before it was decided to return him to Indianapolis as originally planned.

But with Woody Fryman and Dale Murray, acquired from the Expos in the Tony Perez deal, not taking up the slack, Howsam suddenly was working on a complete overhaul of his mound corps.

Eastwick was sent to St. Louis for Doug Capilla, a young lefthander whom the Cincy scouts liked for his strong arm. The ailing Gary Nolan was passed on to the California Angels for a youthful infield prospect. Mike Caldwell, acquired in a spring deal with the Cards, was also sent packing.

The astonishing fact is that only three of the ten pitchers on the 1976 World Series eligibility list are still pitching for the Reds, although a fourth, young Manny Sarmiento, may be back eventually.

Seaver, handsome, articulate and talented, pitched a shutout in Montreal in his debut in Cincinnati uniform. And true to his word Howsam acquired Seaver without disrupting the eight regulars from his championship team. Moskau was called up to the majors but the other four young arms are still on the assembly line.

Nolan, who has come close to bankrupting Blue Cross, was being treated for shoulder problems by his new employers but the question of how much the Reds will miss Eastwick is one that only the future will tell.

Whatever the outcome of the 1977 race, Howsam's boldness will fascinate baseball people for years to come. But it is still true that the Cincinnati team is a baseball force likely to retain its vitality for many years to come.